Mathematical Lectures from Peking University

Editor-in-chief

Gang Tian, Princeton, NJ, USA

More information about this series at http://www.springer.com/series/11574

Michel Broué

On Characters of Finite Groups

 Springer

Michel Broué
Université Paris Diderot Paris 7
Paris
France

ISSN 2197-4209   ISSN 2197-4217 (electronic)
Mathematical Lectures from Peking University
ISBN 978-981-13-4964-5   ISBN 978-981-10-6878-2 (eBook)
https://doi.org/10.1007/978-981-10-6878-2

Mathematics Subject Classification (2010): 20Cxx, 20C15, 16Gxx, 12E99, 05E10

Printed on acid-free paper

This Springer imprint is published by Springer Nature
The registered company is Springer Nature Singapore Pte Ltd.
The registered company address is: 152 Beach Road, #21-01/04 Gateway East, Singapore 189721, Singapore

Preface

This book emerged from a course given twice at Peking University (PKU), each time once a week over two months, while I was visiting the Beijing International Center for Mathematics Research.

I essentially taught character theory of finite groups, insisting during the first year on questions of rationality (hence making no assumption for the ground field to be "big enough"). During the second year, I insisted more on applications, and also on related areas, like questions concerning polynomial invariants of finite groups, or the more recent notion of Drinfeld double of a finite group. Throughout the two courses, I tried systematically—but discretely—to help students become familiar with the language of categories.

The audience consisted of second-year and third-year students. There are not that many places in the world where you could teach such a material in front of undergraduate students… PKU undergraduate students are indeed exceptional: quick, curious, open-minded, just as disbelieving as brilliant students must be, humorous, and obstinate.

Nevertheless, this book contains more than what had actually been taught and, in order to be consistent, it is organized differently from the course. My hope is that its content may be useful for various introductory courses on this fantastic—although not recent—theory. As it is now, the Texte book will usually be better suited for a graduate course (or to exceptionally bright undergraduate students). In exchange for the extra effort, reading this book will give the student a better algebraic formation than reading a book that focusses almost exclusively on groups.

In contrast to some of the recent books on a similar topic (such as [Ste12]), I have chosen not to give classical examples such as representations of the symmetric groups; they can be found in the abundant literature on the subject. Furthermore (unlike in the rather exhaustive treatise [CR87], but also [AB95], or even [Ser12]), I chose not to make use of general "abstract" semisimple algebras: The complete structure theorems about the group algebra (including a Fourier formula over any characteristic zero field, as well as various interpretations of Schur indices) are proven without having to appeal to more general results—only using characters.

This allowed me, in a rather short book, to concentrate on less usual topics, like representations over $\mathbb{Q}$, or the graded G-modules and applications to the action of a group G on $K[X_1, \ldots, X_n]$ if G acts on K^n, or even to present a quick introduction to Hopf algebras and ribbon categories through the study of the Drinfeld double of a finite group, with a natural application to the associated representation of $SL_2(\mathbb{Z})$.

A precision: The first two parts of Jean-Pierre Serre's book [Ser12] have been the main source of inspiration for our treatment of Brauer's Theorem in Chap. 5, as well as for other topics in Chaps. 2 and 3.

Michel me demande de lui faire un dessin pour son nouveau livre. Je lui demande de me raconter "ce qu'il a vu". Je vois passer dans ses yeux l'envie de dire, et l'impossible. Il me dit: "Imagine des objets mathématiques abstraits, très complexes et très harmonieux. Pour les regarder, on les transporte dans un autre monde, un tout petit moins abstrait (qui peut quand même avoir 10 ou 196883 dimensions) et miraculeusement, on les voit beaucoup mieux, leur comportement les révèle." Pendant qu'il parle, ses mains bougent, décrivent des courbes, des vagues, de l'harmonie. Comment dessiner ça ? Je ne vois pas. La seule image qui me vient est celle d'un mathématicien, et plus largement, d'une communauté mathématique, chanceuse de pouvoir naviguer dans tant de dimensions, chanceuse de toucher l'intouchable et d'en revenir les mains pleines. Nous autres, non mathématiciens, ne pouvons qu'admirer ces voyageurs qui prennent le large, qui voient autre chose que le haut et le bas, le concret et l'abstrait. Merci pour ces mystères récoltés dans la solitude.

Anouk Grinberg.

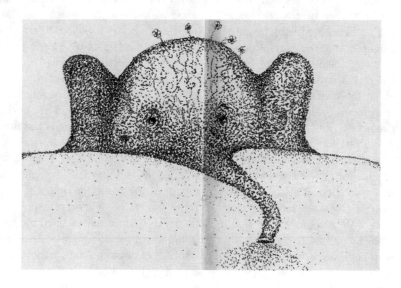

Abstract. The first chapter is devoted to tensor products: This basic and fundamental notion is hardly taught at undergraduate level, and I want the reader to be immediately familiar with it, once and for all.

- Chapter 2 is a general introduction to group representations (using a bit of categorical language). In particular, we treat the case of representations on sets, their classification, as well as Burnside marks, and we also introduce the reader to general linear representations, their language, basic facts such as Schur's lemma, and common examples and counterexamples.
- Chapter 3 contains the general results about characteristic zero representations, without any further assumption about the ground field (hence, the ground field may be $\mathbb{Q}$). For example, it provides a proof of the general "Fourier inversion formula" in this general context, as well as the more classical results about the Galois action on conjugacy classes in relation to the number of irreducible characters. One paragraph is devoted to what was, for Frobenius, the origin of the whole character theory: the group determinant.
- Chapter 4 "plays around with the ground ring." It contains a description, free of the theory of central simple algebras, of the integers which constitute the degree of an irreducible character: Character theory is sufficient to establish, for example, that the dimension of the occurring skewfields over their center is a square. It also contains a short introduction to reflection groups.
- Chapter 5 is devoted to induction–restriction. The first part has no assumption about the characteristic of the field, insisting on the formal equalities and isomorphisms between occurring bimodules or functors. The second part deals, more classically, with the setting of characteristic zero ground fields and class functions.
- The first part of Chap. 6 contains a proof of Brauer's characterization of characters, while its second part contains some (less classical in textbooks) applications to "subgroups controlling π-fusion" and normal π-complements. As an application, it also contains a treatment of Frobenius groups.
- In order to study representations of finite groups on graded modules, Chap. 7 starts with a short introduction to basic notions concerning graded modules and graded algebras. In the second part, we introduce the notion of graded characters with applications to generalization of Molien's Formula. As a particular case, we study the action of G on $S(V)$ (the symmetric algebra of V) if G acts on a vector space V. We conclude by an introduction to the main characterization of reflection groups.
- The last chapter is devoted to the study of the Drinfeld double of a finite group. We start by a quick introduction to Hopf algebras and their algebra representations, then we introduce the notion of universal R-matrix and show that the representation category of the Drinfeld double is a ribbon category. We conclude by the definition of the S-matrix of that ribbon category, and by explicitly computing the associated representation of $GL_2(\mathbb{Z})$.

Prerequisites

This book requires familiarity with the notions of groups, rings, fields, and in particular, with the undergraduate knowledge of linear algebra. More specifically, let k
be a (commutative) field. Here are some examples of what is assumed to be known.

- The results of an undergraduate course on k-linear algebra. If V is a vector space
  over k, we shall denote by $[V : k]$ its dimension, by $\mathrm{tr}_{V/k}(\alpha)$ the trace of an
  endomorphism α of V, and by V^* its dual space.
- Matrices and their determinants.

  For example, the following identity will not be proved: let M be an $n \times n$
  matrix with entries in k, let $^t\mathrm{Com}(M)$ denote the transpose of its matrix of
  cofactors, let 1_n be the $n \times n$ identity matrix; then

$$^t\mathrm{Com}(M) \cdot M = det(M) \cdot 1_n.$$

- Rudiments of fields extensions: multiplicativity of the degrees (if $K \subset L \subset M$
  are fields, then $[M : K] = [M : L][L : K]$), algebraic elements, finite extensions.
- At some places (clearly indicated), which may be omitted by a beginner, we
  shall assume more familiarity with Galois theory, in particular, with Galois
  theory of cyclotomic extensions. All what is needed is treated in the Appendix.

We take for granted that the reader is familiar with the standard notation $\mathbb{N}$ (for
"numbers")—note that by convention $\mathbb{N} = \{0, 1, 2, \ldots\}$, $\mathbb{Z}$ (for "Zahlen"), $\mathbb{Q}$ (for
"quotients"), $\mathbb{R}$ (for "real"), $\mathbb{C}$ (for "complex"), as well as $\mathbb{F}_p = \mathbb{Z}/p\mathbb{Z}$ (for "finite").
Moreover, the letter i will denote a complex number such that $i^2 = -1$.

By convention, a *division ring* is a (not necessarily commutative) ring where all
nonzero elements are invertible—a division ring is also called a skewfield. A *field* is
a *commutative* division ring.

The letter k (for "Körper") will usually denote a field; K, L will usually denote
characteristic zero fields[1], while D will be used for division rings.

The center ZD of a division ring D is a field, and if k is a subfield of ZD, we say
that D is a division k-algebra.

For D a division ring and V a D-vector space, we denote by $[V : D]$ the
dimension of V. In particular, if D is a division k-algebra for a field k, $[D : k]$ is the
dimension of D viewed as a k-vector space. Only in the last chapter, where the field
k will be fixed, will we denote the dimension of a k-vector space V by $dim\ V$.

---

[1] It may happen also that L denotes a subgroup of a group $G\ldots$

We shall also use the following notation.

- The *Kronecker delta function*, applied to two variables x and y, is denoted $\delta_{x,y}$—not to be confused with the characteristic functions δ_s (unbolded) used in the last chapter.
- For Ω any finite set, $|\Omega|$ will denote the number of its elements.
- A subset (subgroup, subring, submodule, …) Ω' of a set (group, ring, module, …) Ω is said to be *proper* if $\Omega' \neq \Omega$.
- If g is an element of a group G, we denote by g the subgroup generated by g.
- For H a subgroup of a group G, we denote by $[G/H]$ a complete set of representatives for the cosets gH of G modulo H.
- For a group G acting on a set Ω, if $\mathrm{Cl}(\Omega)$ is the set of orbits of Ω under G, we shall denote by $[\mathrm{Cl}(\Omega)]$ a complete set of representatives for the orbits.

Paris, France

Michel Broué

Acknowledgements

I warmly thank Gunter Malle for his careful and patient readings and corrections of the manuscript. My thanks go also to the students who attended the courses which motivated these notes, for their interest, their attention, their questions, their wonder for Mathematics.

Contents

Chapter 1
Tensor Product

Throughout this chapter, k is a field.

1.1 Definition of the Tensor Product

In what follows, M and N are k-vector spaces.

The tensor product of M and N *"linearizes the bilinear maps"* in the following sense.

Definition 1.1.1 A *tensor product of M and N* is a pair (T, t) where

- T is a k-vector space,
- $t : M \times N \to T$ is a bilinear map,

which satisfies the following "universal property":

Whenever X is a k-vector space and $f : M \times N \to X$ is a bilinear map, there exists a unique k-linear map $\overline{f} : T \to X$ such that the diagram

$$
\begin{array}{ccc}
M \times N & \xrightarrow{\ f\ } & X \\
{\scriptstyle t}\downarrow & \nearrow{\scriptstyle \overline{f}} & \\
T & &
\end{array}
$$

is commutative.

As the reader can understand, the next result may be viewed as a routine argument for "universal properties" like the one described in Definition 1.1.1 above.

© Springer Nature Singapore Pte Ltd. 2017
M. Broué, *On Characters of Finite Groups*, Mathematical Lectures from Peking University, https://doi.org/10.1007/978-981-10-6878-2_1

Lemma 1.1.2 *If a tensor product* (T, t) *exists, then it is unique, in the following sense.*

If (T', t') *is another tensor product, then there is a unique isomorphism*

$$\phi : T \xrightarrow{\sim} T'$$

such that the following diagram commutes

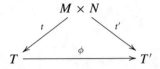

Proof First apply the universal property to the tensor product (T, t) with $X := T'$ and $f = t'$. We get a unique linear map $\phi : T \xrightarrow{\sim} T'$ which makes the above diagram commutative.

Then apply the universal property to the tensor product (T', t') with $X := T$ and $f = t$. We get a unique linear map $\phi' : T' \xrightarrow{\sim} T$ which makes the following diagram commutative:

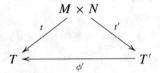

Finally apply the universal property to the tensor product (T, t) with $X := T$ and $f = t$. We see that $\phi' \cdot \phi = \mathrm{Id}_T$. Similarly, we get $\phi \cdot \phi' = \mathrm{Id}_{T'}$. Thus ϕ is indeed an isomorphism.   □

We shall prove now that the tensor product of vector spaces exists.

Remark 1.1.3 Since the tensor product is defined up to unique isomorphism, it is enough to provide a construction – then the construction may be forgotten, only keeping in mind that it does exist.

Let us consider the module $k^{M \times N}$ – that is, the k-vector space with a basis $(\mathbf{e}_{(m,n)})$ indexed by the elements of $M \times N$. Let B be its subspace generated by the following set:

$$\left\{ \mathbf{e}_{(m_1+m_2,n)} - \mathbf{e}_{(m_1,n)} - \mathbf{e}_{(m_2,n)} \, , \; \mathbf{e}_{(m,n_1+n_2)} - \mathbf{e}_{(m,n_1)} - \mathbf{e}_{(m,n_2)} \, , \right.$$
$$\left. \lambda \mathbf{e}_{(m,n)} - \mathbf{e}_{(\lambda m,n)} \, , \; \lambda \mathbf{e}_{(m,n)} - \mathbf{e}_{(m,\lambda n)} \right\}_{m,m_1,m_2 \in M \, , \; n,n_1,n_2 \in N \, , \; \lambda \in k}$$

We define

$$\begin{cases} M \otimes_k N := k^{M \times N}/B \,, \\ \iota_{M,N} : M \times N \to M \otimes_k N \,, \quad (m, n) \mapsto \pi_B(\mathbf{e}_{(m,n)}) \,, \end{cases}$$

where $\pi_B : k^{M \times N} \twoheadrightarrow M \otimes_k N$ denotes the natural surjection.

It is clear (isn't it?) that the map $\iota_{M,N}$ is well defined and is bilinear. Notice that the space $M \otimes_k N$ is generated by the family $\iota_{M,N}(m, n)$ for $m \in M$ and $n \in N$.

Now if X is a k-vector space and $f : M \times N \to X$ is a bilinear map, the linear map $k^{M \times N} \to X$ induced by f vanishes on B, hence induces a linear map $\overline{f} : M \otimes_k N \to X$. The commutativity of the diagram is clear, and the unicity of $\overline{f}$ is a consequence of the fact that $M \otimes_k N$ is generated by the family $\iota_{M,N}(m, n)$ for $m \in M$ and $n \in N$.

Notation 1.1.4   From now on, for $m \in M$ and $n \in N$, we set

$$m \otimes_k n := \iota_{M,N}(m, n)$$

and the elements $m \otimes_k n$ are called the *elementary tensors*.

$\bigodot$ For M and N two vector spaces of dimension at least 2, by no way is the family $(m \otimes_k n)_{m \in M, n \in N}$ the set of all elements of $M \otimes_k N$ nor a basis of $M \otimes_k N$, even if M and N are finite dimensional vector spaces — see for example Proposition 1.2.14 and Remark 1.2.15 below.

Let $\mathrm{Bil}(M, N; X)$ denote the space of all bilinear maps from $M \times N$ to X. By Definition 1.1.1, the map $f \mapsto \overline{f}$ is a "natural" isomorphism ("linearization of the bilinearity")

$$\mathrm{Bil}(M, N; X) \overset{\sim}{\longrightarrow} \mathrm{Hom}(M \otimes_k N, X)\,.$$

Remark 1.1.5   In particular, we have a natural isomorphism

$$\mathrm{Bil}(M, N; k) \overset{\sim}{\longrightarrow} \mathrm{Hom}(M \otimes_k N, k) = (M \otimes_k N)^*\,.$$

It is well known that we also have a natural isomorphism

$$\begin{cases} \mathrm{Bil}(M, N; X) \overset{\sim}{\longrightarrow} \mathrm{Hom}(M, \mathrm{Hom}(N, X)) \\ f \mapsto \hat{f} \quad \text{where } \hat{f} : m \mapsto (n \mapsto f(m, n))\,. \end{cases}$$

Thus we get

Proposition 1.1.6   *The map* $\overline{f} \mapsto \hat{f}$ *is an isomorphim*

$$\mathrm{Hom}(M \otimes_k N, X) \overset{\sim}{\longrightarrow} \mathrm{Hom}(M, \mathrm{Hom}(N, X))\,.$$

1.2   Properties of the Tensor Product

1.2.1   Functoriality

The next proposition expresses in particular the fact that the tensor product is a *bifunctor*.

Proposition 1.2.1 *Suppose given linear maps* $\alpha : M \to M'$ *and* $\beta : N \to N'$ *between k-vector spaces.*

(1)   Then there exists a unique linear map

$$\alpha \otimes_k \beta : M \otimes_k N \to M' \otimes_k N'$$

such that, for all $m \in M$ *and* $n \in N$,

$$(\alpha \otimes_k \beta)(m \otimes_k n) = \alpha(m) \otimes_k \beta(n).$$

(2) $\mathrm{Id}_M \otimes_k \mathrm{Id}_N = \mathrm{Id}_{M \otimes_k N}$.
(3)   Given linear maps $\alpha' : M' \to M''$ *and* $\beta' : N' \to N''$, *we have*

$$(\alpha' \otimes_k \beta') \circ (\alpha \otimes_k \beta) = (\alpha' \circ \alpha) \otimes_k (\beta' \circ \beta).$$

Sketch of proof (1) follows from the universal property applied to the diagonal map in the following commutative diagram:

$$
\begin{array}{ccc}
M \times N & \xrightarrow{\ \ \alpha \times \beta\ \ } & M' \times N' \\
{\scriptstyle t_{M,N}}\downarrow & & \downarrow{\scriptstyle t_{M',N'}} \\
M \otimes_k N & \dashrightarrow{\ \ \alpha \otimes_k \beta\ \ } & M' \otimes_k N'.
\end{array}
$$

(2) is trivial and (3) follows from the unicity in the universal property.   □

1.2.2   More Properties

In what follows, linear maps on tensor products are defined by their values on elementary tensors, according to the universal property.

Lemma 1.2.2 *Let M be a k-vector space.*

(1) $k \otimes_k M = \{1 \otimes_k m \mid m \in M\}$.
(2)   The maps

$$\begin{cases} k \otimes_k M \to M \, , \ \lambda \otimes_k m \mapsto \lambda m \, , \\ M \to k \otimes_k M \, , \ m \mapsto 1 \otimes_k m \, , \end{cases}$$

are mutually inverse isomorphisms.

Proof (1) follows from the equality $\lambda \otimes_k m = 1 \otimes_k \lambda m$, from the formula $\sum_i 1 \otimes_k m_i = 1 \otimes_k \sum_i m_i$, and from the fact that $k \otimes_k M$ is generated by the elementary tensors.

(2) is trivial.

□

Proposition 1.2.3 *Let M and N be k-vector spaces.*

(1) Whenever $m \in M$ and $n \in N$, $m \otimes_k n = 0$ implies $m = 0$ or $n = 0$.
(2) Whenever $m \in M$, $m \neq 0$, the map

$$N \to M \otimes_k N \, , \ n \mapsto m \otimes_k n \, ,$$

is an injective k-linear map.
(3) In particular its image

$$m \otimes_k N := \{ m \otimes_k n \mid n \in N \}$$

is a subspace of $M \otimes_k N$ isomorphic to N.

Proof Notice first that $m \otimes_k N = km \otimes_k N$. Since the map $\lambda \mapsto \lambda m$ is an isomorphism of k-vector spaces $k \xrightarrow{\sim} km$, it follows from the functoriality (see Proposition 1.2.1 above) that the map

$$\begin{cases} m \otimes_k N \to N \, , \\ m \otimes_k n \mapsto n \end{cases}$$

is an isomorphism, proving both (1) and (2) (hence (3)).

□

The proofs of the statements of the following proposition are left as exercises to the reader.

Proposition 1.2.4 (Functorial isomorphisms)

(1) We have unique isomorphisms

$$\begin{cases} (M_1 \otimes_k M_2) \otimes_k M_3 \xrightarrow{\sim} M_1 \otimes_k (M_2 \otimes_k M_3) \, , \\ (m_1 \otimes_k m_2) \otimes_k m_3 \mapsto m_1 \otimes_k (m_2 \otimes_k m_3) \, , \end{cases}$$

$$\begin{cases} M_1 \otimes_k M_2 \xrightarrow{\sim} M_2 \otimes_k M_1 \, , \\ m_1 \otimes_k m_2 \mapsto m_2 \otimes_k m_1 \, , \end{cases}$$

and they are functorial, i.e., given linear maps $\alpha_i : M_i \to M_i'$ (for $i = 1, 2, 3$), the following diagrams commute

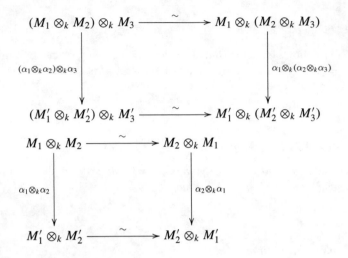

(2)   as well as

$$\begin{cases} M_1 \otimes_k (M_2 \oplus M_3) \xrightarrow{\sim} (M_1 \otimes_k M_2) \oplus (M_1 \otimes_k M_3)\,, \\ m_1 \otimes_k (m_2 + m_3) \mapsto (m_1 \otimes_k m_2) + (m_1 \otimes_k m_3)\,, \end{cases}$$

and more generally

$$\begin{cases} M \otimes_k \left(\bigoplus_{i \in I} M_i \right) \xrightarrow{\sim} \bigoplus_{i \in I} (M \otimes_k M_i)\,, \\ m \otimes_k \left(\sum_{i \in I} m_i \right) \mapsto \sum_{i \in I} (m \otimes_k m_i)\,, \end{cases}$$

and they are functorial, i.e., given linear maps $\alpha_i : M_i \to M_i'$ (for $i = 1, 2, 3$), the following diagram commutes

$$
\begin{array}{ccc}
M_1 \otimes_k (M_2 \oplus M_3) & \xrightarrow{\sim} & (M_1 \otimes_k M_2) \oplus (M_1 \otimes_k M_3) \\
\big\downarrow{\scriptstyle \alpha_1 \otimes_k (\alpha_2 \oplus \alpha_3)} & & \big\downarrow{\scriptstyle (\alpha_1 \otimes_k \alpha_2) \oplus (\alpha_1 \otimes_k \alpha_3)} \\
M_1' \otimes_k (M_2' \oplus M_3') & \xrightarrow{\sim} & (M_1' \otimes_k M_2') \oplus (M_1' \otimes_k M_3')
\end{array}
$$

(3)   as well as (for any set I)

$$\begin{cases} k^{(I)} \otimes_k M \xrightarrow{\sim} M^{(I)} , \\ (\lambda_i)_{i \in I} \otimes_k m \mapsto (\lambda_i m)_{i \in I} , \end{cases}$$

and it is functorial (statement left to the reader).

The next proposition holds in full generality, even if M or N are not finite dimensional.

Proposition 1.2.5 *(1) Assume that M has basis $(e_i)_{i \in I}$. Then we have*

$$M \otimes_k N = \bigoplus_{i \in I} e_i \otimes_k N .$$

(2) Assume moreover that N has basis $(f_j)_{j \in J}$. Then $M \otimes_k N$ has basis $(e_i \otimes_k f_j)_{i \in I, j \in J}$.
(3) We have $\dim(M \otimes_k N) = \dim(M) \dim(N)$.

Sketch of proof It is an easy consequence of item (3) of Proposition 1.2.4. □

Corollary 1.2.6 *The map*

$$k[X] \otimes_k k[Y] \to k[X, Y] , \quad P(X) \otimes_k Q(Y) \mapsto P(X)Q(Y)$$

is an isomorphism of k-vector spaces.

Proof A basis of $k[X]$ is $(X^m)_{m \in \mathbb{N}}$, a basis of $k[Y]$ is $(Y^n)_{n \in \mathbb{N}}$, and a basis of $k[X, Y]$ is $(X^m Y^n)_{m, n \in \mathbb{N}}$. Now apply Proposition 1.2.5. □

Exercise 1.2.7 One may notice that

- the vector space $k[X] \otimes_k k[Y]$ is naturally endowed with an algebra structure, such that $(x \otimes_k y)(x' \otimes_k y') = (xx' \otimes_k yy')$ (for $x, x' \in k[X]$ and $y, y' \in k[Y]$),
- and that the map described in Corollary 1.2.6 is an algebra isomorphism.

Prove that the algebra $k(X) \otimes_k k(Y)$ (defined as above) is *not* a field, hence that $k(X) \otimes_k k(Y)$ is not isomorphic (as a k-algebra) to $k(X, Y)$.

HINT. One can prove that if K and K' are (commutative) fields extensions of a field k and if there exists a field isomorphism $\sigma : K \xrightarrow{\sim} K'$ which induces the identity on k, then the k-algebra $K \otimes_k K'$ is not a field.

1.2.3 Kronecker Product of Matrices

Let $p, q \in \mathbb{N}$, set $I := \{1, \ldots, p\}$ and $J = \{1, \ldots, q\}$, and let $A = (a_{i_1, i_2})_{i_1, i_2 \in I}$ and $B = (b_{j_1, j_2})_{j_1, j_2 \in J}$ be matrices over k.

Definition 1.2.8 The Kronecker product of A and B is the matrix denoted $A \otimes B$ indexed by $I \times J$ (endowed with the lexicographic order) with entries

$$c_{(i_1,j_1),(i_2,j_2)} := a_{i_1,i_2} b_{j_1,j_2} \,.$$

Remark 1.2.9 A mnemonic way to remember that definition is to write

$$A \otimes B = \begin{pmatrix} a_{1,1}B & \cdots & a_{1,p}B \\ \vdots & \ddots & \vdots \\ a_{p,1}B & \cdots & a_{p,p}B \end{pmatrix}.$$

Note that the matrix

$$\begin{pmatrix} Ab_{1,1} & \cdots & Ab_{1,q} \\ \vdots & \ddots & \vdots \\ Ab_{q,1} & \cdots & Ab_{q,q} \end{pmatrix}$$

corresponds to another order on $I \times J$ (which one?).

Example 1.2.10 Assume that $p = q = 2$, and

$$A = \begin{pmatrix} a_{1,1} & a_{1,2} \\ a_{2,1} & a_{2,2} \end{pmatrix} , \quad B = \begin{pmatrix} b_{1,1} & b_{1,2} \\ b_{2,1} & b_{2,2} \end{pmatrix}.$$

Then

$$A \otimes B = \begin{pmatrix} a_{1,1}b_{1,1} & a_{1,1}b_{1,2} & a_{1,2}b_{1,1} & a_{1,2}b_{1,2} \\ a_{1,1}b_{2,1} & a_{1,1}b_{2,2} & a_{1,2}b_{2,1} & a_{1,2}b_{2,2} \\ a_{2,1}b_{1,1} & a_{2,1}b_{1,2} & a_{2,2}b_{1,1} & a_{2,2}b_{1,2} \\ a_{2,1}b_{2,1} & a_{2,1}b_{2,2} & a_{2,2}b_{2,1} & a_{2,2}b_{2,2} \end{pmatrix}.$$

Assume M and N are finite dimensional k-vector spaces, of dimensions respectively p and q. Let $(e_i)_{i \in I}$ and $(f_j)_{j \in J}$ be respectively bases of M and N.

Let α and β be endomorphisms of respectively M and N, and let us denote by $A = (a_{i_1,i_2})$ and $B = (b_{j_1,j_2})$ their matrices on the above bases.

The following property of the Kronecker product is an immediate consequence of the above notation and of the definition of $\alpha \otimes_k \beta$.

Proposition 1.2.11 *With the above notation and conventions, the Kronecker product $A \otimes B$ of A and B is the matrix of $\alpha \otimes_k \beta$ on the basis $(e_i \otimes_k f_j)_{(i,j) \in I \times J}$ of $M \otimes_k N$.*

Corollary 1.2.12 *Let M and N be finite dimensional k-vector spaces, and let α and β be endomorphisms of respectively M and N. Then*

$$\mathrm{tr}(\alpha \otimes_k \beta) = \mathrm{tr}(\alpha)\mathrm{tr}(\beta) \,.$$

Proof Choosing bases for M and N as above, and using notation from Definition 1.2.8, we see that

$$\text{tr}(\alpha \otimes_k \beta) = \sum_{i,j} c_{(i,j),(i,j)} = \sum_{i,j} a_{i,i} b_{j,j} = \text{tr}(\alpha)\text{tr}(\beta) \,.$$

$\square$

Exercise 1.2.13 (1) Prove that $\det(A \otimes B) = \det(A)^q \det(B)^p$.
(2) Prove that $\text{rk}(A \otimes B) = \text{rk}(A)\text{rk}(B)$.

1.2.4 Tensor Product and Homomorphisms

Proposition 1.2.14 *Let M and N be finite dimensional k-vector spaces.*

(1) The bilinear map

$$\begin{cases} M^* \times N \longrightarrow \text{Hom}_k(M, N) \,, \\ (\mu, n) \mapsto (m \mapsto \mu(m)n) \,, \end{cases}$$

induces an isomorphism

$$M^* \otimes_k N \overset{\sim}{\longrightarrow} \text{Hom}_k(M, N) \,.$$

(2) If $(e_i)_{i \in I}$ is a basis of M, and if $(e_i^*)_{i \in I}$ denotes the dual basis (a basis of M^*), the inverse of the above isomorphism is

$$\begin{cases} \text{Hom}_k(M, N) \longrightarrow M^* \otimes_k N \,, \\ \phi \mapsto \sum_{i \in I} e_i^* \otimes_k \phi(e_i) \,. \end{cases}$$

(3) For $N = M$, the composition of the linear map

$$M^* \otimes_k M \to k \,, \quad \mu \otimes_k m \mapsto \mu(m)$$

with the inverse of the above isomorphism is the trace map

$$\text{tr} : \text{End}_k(M) \to k \,.$$

Proof (1) The images of the elementary tensors under the described morphism $M^* \otimes_k N \to \text{Hom}_k(M, N)$ are exactly all the linear maps from M to N of rank one. Since those maps generate $\text{Hom}_k(M, N)$, this implies that the morphism $M^* \otimes_k N \to$

$\text{Hom}_k(M, N)$ is onto. Since both sides have the same dimension, that proves that it is an isomorphism.

Remark 1.2.15 Through the preceding isomorphism, the nonzero elementary tensors correspond to the rank one linear maps from M to N.

(2) The image of ϕ under the composition

$$\text{Hom}_k(M, N) \to M^* \otimes_k N \to \text{Hom}_k(M, N)$$

is the map

$$x \mapsto \sum_i e_i^*(x)\phi(e_i) = \phi\left(\sum_i e_i^*(x)(e_i)\right) = \phi(x).$$

The image of $\mu \otimes_k n$ under the composition

$$M^* \otimes_k N \to \text{Hom}_k(M, N) \to M^* \otimes_k N$$

is

$$\sum_i e_i^* \otimes_k \mu(e_i)n = \sum_i \mu(e_i)e_i^* \otimes_k n = \left(\sum_i \mu(e_i)e_i^*\right) \otimes_k n = \mu \otimes_k n.$$

(3) The composition of the linear map

$$M^* \otimes_k M \to k , \quad \mu \otimes_k m \mapsto \mu(m)$$

with the inverse of the isomorphism $M^* \otimes_k M \xrightarrow{\sim} \text{End}_k(M)$ coincides with the trace map on linear maps of rank one. Hence it is the trace. $\qquad\square$

Exercise 1.2.16 Use Proposition 1.2.14 to provide another proof (not using matrices) of Corollary 1.2.12.

1.2.5   Extension of Scalars

The next proposition allows us to "extend the scalars" of vector spaces.

Proposition 1.2.17 *Assume that k is a subfield of a field k'; thus k' is naturally a k-vector space. Let M be a k-vector space. Then the k-bilinear map*

$$\begin{cases} k' \times (k' \otimes_k M) \to k' \otimes_k M \\ (\lambda_1', \lambda_2' \otimes_k m) \mapsto \lambda_1'\lambda_2' \otimes_k m \end{cases}$$

defines a structure of k'-vector space on the k-vector space $k' \otimes_k M$.

The following properties are straightforward.

Proposition 1.2.18 *(1) If $(e_i)_{i \in I}$ is a basis of the k-vector space M, then $(1 \otimes_k e_i)_{i \in I}$ is a basis of the k'-vector space $k' \otimes_k M$.*
(2) Extending scalars is functorial: given a linear map $\phi : M \to N$, the map

$$\mathrm{Id}_{k'} \otimes_k \phi : k' \otimes_k M \to k' \otimes_k N$$

is a k'-linear map, such that

(a) for $\phi_1 : M_1 \to M_2$ and $\phi_2 : M_2 \to M_3$ linear maps, we have

$$\mathrm{Id}_{k'} \otimes_k (\phi_2 \circ \phi_1) = (\mathrm{Id}_{k'} \otimes_k \phi_2) \circ (\mathrm{Id}_{k'} \otimes_k \phi_1).$$

(b) $\mathrm{Id}_{k'} \otimes_k \mathrm{Id}_M = \mathrm{Id}_{k' \otimes_k M}$.

(3) With the same notation as above, if M and N are finite dimensional with bases respectively $(e_i)_{i \in I}$ and $(f_j)_{j \in J}$, and if $\phi : M \to N$ is linear, the matrix of $\mathrm{Id}_{k'} \otimes_k \phi$ on the bases $(1 \otimes_k e_i)_{i \in I}$ and $(1 \otimes_k f_j)_{j \in J}$ equals the matrix of ϕ on the bases $(e_i)_{i \in I}$ and $(f_j)_{j \in J}$.

1.2.6 Trace and Restriction of Scalars

Lemma 1.2.19 *Let k'/k be a finite field extension, let M be a finite dimensional k'-vector space, and let $\alpha \in \mathrm{End}_{k'}(M)$. Then M may be viewed as a k-vector space, α may be viewed as an endomorphism of that k-vector space, with trace denoted by $\mathrm{tr}_{M/k}(\alpha)$. We have*

$$\mathrm{tr}_{M/k}(\alpha) = \mathrm{tr}_{k'/k}(\mathrm{tr}_{M/k'}(\alpha)).$$

Proof of lemma 1.2.19. It is enough to establish the formula when α has rank one, i.e., when α corresponds to an elementary tensor $\mu \otimes_{k'} m$ under the isomorphism $M^* \otimes_{k'} M \xrightarrow{\sim} \mathrm{End}_{k'}(M)$ (see Proposition 1.2.14). In that case,

$$\alpha(x) = \mu(x)m \text{ and } \mathrm{tr}_{M/k'}(\alpha) = \mu(m).$$

Let $(e_i)_{i \in I}$ be a basis of k' over k, and let $(e_i^*)_{i \in I}$ be its dual basis. Then whenever $\lambda \in k'$, $\lambda = \sum_{i \in I} e_i^*(\lambda)e_i$.

- Thus $\mathrm{tr}_{k'/k}(\mathrm{tr}_{M/k'}(\alpha))$ is the trace of the endomorphism of k' defined by $e_i \mapsto e_i \mu(m) = \mu(e_i m)$ (for $i \in I$) since μ is k'-linear. It follows that this endomorphism of k' corresponds to the element

$$\sum_{i \in I} e_i^* \otimes_k \mu(e_i m) \in \text{Hom}_k(k', k) \otimes_k k',$$

and its trace is

$$\text{tr}_{k'/k}(\text{tr}_{M/k'}(\alpha)) = \sum_{i \in I} e_i^*(\mu(e_i m)).$$

• For $x \in M$, we have $\mu(x) = \sum_{i \in I}(e_i^* \mu)(x)e_i$, which implies that $\alpha(x) = \sum_{i \in I}(e_i^* \mu)(x)e_i m$, and so α corresponds to the element

$$\sum_{i \in I}(e_i^* \mu) \otimes_k e_i m \in \text{Hom}_k(M, k) \otimes_k M,$$

hence

$$\text{tr}_{M/k}(\alpha) = \sum_{i \in I}(e_i^* \mu)(e_i m).$$

The lemma is now obvious.   □

1.3 Symmetric and Alternating Powers

1.3.1 Symmetric and Alternating Squares

Assume k has characteristic different from 2.

For M a finite-dimensional k-vector space, consider (see Proposition 1.2.4, (1)) the automorphism (called the *flip* Φ)

$$\Phi : \begin{cases} M \otimes_k M \to M \otimes_k M \\ m_1 \otimes_k m_2 \mapsto m_2 \otimes_k m_1. \end{cases}$$

Since $\Phi^2 = \Phi$, we have

$$M \otimes_k M = \ker(\Phi - 1) \oplus \ker(\Phi + 1).$$

Definition 1.3.1 We set

• $\text{Sym}^2(M) := \ker(\Phi - 1)$,
• $\text{Alt}^2(M) := \ker(\Phi + 1)$.

Thus

$$M \otimes_k M = \mathrm{Sym}^2(M) \oplus \mathrm{Alt}^2(M).$$

Proposition 1.3.2 *Let $(e_i)_{i=1,...,r}$ be a basis of M.*

- *The family $(e_i \otimes_k e_j + e_j \otimes_k e_i)_{i \leq j}$ is a basis of $\mathrm{Sym}^2(M)$. Hence*

$$\dim(\mathrm{Sym}^2(M)) = \binom{n}{2} + n = \frac{n(n+1)}{2}.$$

- *The family $(e_i \otimes_k e_j - e_j \otimes_k e_i)_{i < j}$ is a basis of $\mathrm{Alt}^2(M)$. Hence*

$$\dim(\mathrm{Alt}^2(M)) = \binom{n}{2} = \frac{n(n-1)}{2}.$$

1.3.2 Tensor, Symmetric and Exterior Algebras

Throughout this paragraph, M denotes a k-vector space with finite dimension r.
  Tensor algebra.
  We let the reader check the following properties.

Proposition 1.3.3 *(1) For any natural integer $n \geq 0$, one may define*

- *a vector space $M^{\otimes^n} := M \otimes_k \cdots \otimes_k M$ (n factors M),*
- *together with a multilinear map*

$$\begin{cases} M \times \cdots \times M \to M^{\otimes^n}, \\ (v_1, \ldots, v_n) \mapsto v_1 \otimes \cdots \otimes v_n, \end{cases}$$

which is uniquely defined by the following universal property.
(2) Given any k-vector space X and any multilinear map $f : M \times \cdots \times M \to X$, there exists a unique k-linear map $\widehat{f} : M^{\otimes^n} \to X$ such that

$$f(v_1, \ldots, v_n) = \widehat{f}(v_1 \otimes \cdots \otimes v_n).$$

(3) If m and n are two natural integers, there is a unique bilinear map

$$\begin{cases} M^{\otimes^m} \times M^{\otimes^n} \to M^{\otimes^{m+n}}, \\ (v_1 \otimes \cdots \otimes v_m, \, v_{m+1} \otimes \cdots \otimes v_{m+n}) \mapsto v_1 \otimes \cdots \otimes v_{m+n}. \end{cases}$$

Definition 1.3.4 The tensor algebra of M is the k-vector space

$$T(M) := k \oplus M \oplus (M \otimes M) \oplus \cdots \oplus M^{\otimes^n} \oplus \cdots$$

endowed with the multiplication such that

$$(v_1 \otimes \cdots \otimes v_m) \cdot (v_{m+1} \otimes \cdots \otimes v_{m+n}) := v_1 \otimes \cdots \otimes v_{m+n} .$$

Exercise 1.3.5 Check that this does define a structure of k-algebra on $T(M)$.

We denote by $\iota_M : M \hookrightarrow T(M)$ the natural injection.
The proof of the following theorem is immediate.

Theorem 1.3.6 [UNIVERSAL PROPERTY OF $(T(M), \iota_M)$] *Given a k-algebra A and a k-linear map $f : M \to A$, there is a unique algebra morphism $\widehat{f} : T(M) \to A$ such that the diagram*

commutes.

Symmetric algebra.

Definition 1.3.7 The symmetric algebra $S(M)$ is the quotient of $T(M)$ by the twosided ideal generated by

$$\{v_1 \otimes v_2 - v_2 \otimes v_1 \mid v_1, v_2 \in M\},$$

endowed with the natural injection $\iota_M : M \hookrightarrow S(M)$ (why is it an injection?).

Theorem 1.3.8 *(1)* [UNIVERSAL PROPERTY OF $(S(M), \iota_M)$] *Whenever A is a k-algebra and $f : M \to A$ is a linear map such that $f(v)f(v') = f(v')f(v)$ for all $v, v' \in M$, there exists a unique k-algebra morphism $\widehat{f} : S(M) \to A$ such that the diagram*

commutes.
(2) Let us choose a basis $(e_1, \ldots, e_r)$ of M.

(a) *There is a unique algebra morphism*

$$p_M : \begin{cases} S(M) \to k[X_1, \ldots, X_r] \\ e_j \mapsto X_j . \end{cases}$$

(b) *The above algebra morphism is an isomorphism.*

Remark 1.3.9 As shown by item (2) of the above theorem, the symmetric algebra $S(M)$ should be seen as playing for the polynomial algebra $k[X_1, \ldots, X_r]$ the same role as the vector space M is playing for K^r: an "intrinsic" version avoiding the choice of a basis.

Exterior algebra.

Definition 1.3.10 The exterior algebra $\Lambda(M)$ is the quotient of $T(M)$ by the twosided ideal generated by

$$\{v \otimes v \mid v \in M\},$$

endowed with the natural injection $\iota_M : M \hookrightarrow \Lambda(M)$ (why is it an injection?).

If $x, y \in \Lambda(M)$, we denote by $x \wedge y$ their product.

Theorem 1.3.11 [UNIVERSAL PROPERTY OF $(\Lambda(M), \iota_M)$] *Whenever A is a k-algebra and $f : M \to A$ is a linear map such that $f(v)^2 = 0$ for all $v \in M$, there exists a unique k-algebra morphism $\widehat{f} : \Lambda(M) \to A$ such that the diagram*

commutes.

1.4 Tensor Product over an Algebra

Left and right modules.
Let A be a k-algebra.

An A-module (or "left A-module") is a k-vector space M endowed with a k-algebra morphism $A \to \text{End}_k(M)$.

Thus an A-module M is endowed with a (left) multiplication by the elements of A: for $a \in A$ and $m \in M$, the map

$$M \to M , \quad m \mapsto am$$

is k-linear and is associative: $(ab)m = a(bm)$ for all $a, b \in A$ and $m \in M$.

A module-A M (or "right A-module") is a k-vector space M endowed with a k-algebra anti-morphism $A \to \operatorname{End}_k(M)$, that is M is endowed with a right multiplication by the elements of A: for $a \in A$ and $m \in M$, the map

$$M \to M \ , \ m \mapsto ma$$

is k-linear and is associative: $m(ab) = (ma)b$ for all $a, b \in A$ and $m \in M$.

Tensor product of left and right modules.

Definition 1.4.1 Let M be a module-A and let N be an A-module.

(1) We set

$$M \otimes_A N := M \otimes_k N \Big/ \Big\{ \sum_{\substack{a \in A \\ m \in M, n \in N}} (ma \otimes_k n - m \otimes_k an) \Big\}.$$

(2) For $m \in M$ and $n \in N$ we denote by $m \otimes_A n$ the image of $m \otimes_k n$ in $M \otimes_A N$.

Notice that the map

$$M \times N \to M \otimes_A N \ , \ (m, n) \mapsto m \otimes_A n$$

is k-bilinear and that, for all $a \in A, m \in M, n \in N$,

$$ma \otimes_A n = m \otimes_A an \ .$$

The following expresses a universal property. Its proof is easy and is left to the reader.

Proposition 1.4.2 *Let M be a module-A and let N be an A-module. Whenever*

- *X is a k-vector space,*
- *$f : M \times N \to X$ is a k-bilinear map such that,*

$$f(ma, n) = f(m, an)$$

for all $a \in A, m \in M, n \in N$,

there exists a unique k-linear map $\widehat{f} : M \otimes_A N \to X$ such that

$$\widehat{f}(m \otimes_A n) = f(m, n)$$

for all $m \in M, n \in N$.

Tensor product and direct sums.

Proposition 1.4.3 *Assume that a module-A is a direct sum of submodules-A:*

$$M = \bigoplus_{i \in I} M_i \,.$$

Whenever N is an A-module,

$$M \otimes_A N = \bigoplus_{i \in I} M_i \otimes_A N \,.$$

Exercise 1.4.4 Prove the above proposition.

Tensor product and bimodules.

Let A and B be k-algebras. An *A-module-B* (also called (A, B)-bimodule) is a k-vector space M
- which is both an A-module and a module-B,
- in such a way that, for all $a \in A$, $b \in B$, $m \in M$,

$$a(mb) = (am)b \,.$$

Proposition 1.4.5 *Let M be an A-module-B.*

(1) Let Y be a B-module. The k-vector space $M \otimes_B Y$ is naturally endowed with a structure of A-module, defined by the condition

$$a(m \otimes_B y) := am \otimes_B y$$

for all $a \in A$, $y \in Y$, $m \in M$.
(2) Let $\beta : Y \rightarrow Y'$ be a morphism of B-modules. Then there is a (unique) morphism of A-modules

$$\mathrm{Id}_M \otimes_B \beta : M \otimes_B Y \rightarrow M \otimes_B Y' \,, \quad m \otimes_B y \mapsto m \otimes_B \beta(y) \,.$$

Exercise 1.4.6 Prove the above proposition.

More Exercises on Chap. 1

Exercise 1.4.7 Let M and N be k-vector spaces.

(1) Describe a natural nonzero linear map

$$M^* \otimes_k N^* \to (M \otimes_k N)^* .$$

(2) Assume that M and N are *finite dimensional*. Prove that the above map is an isomorphism.
(3) Assuming again M and N finite dimensional, prove that there is a natural isomorphism

$$\text{End}(M) \otimes_k \text{End}(N) \xrightarrow{\sim} \text{End}(M \otimes_k N) .$$

Exercise 1.4.8 Let M and N be finite dimensional k-vector spaces.

(1) Let $m_1, m_2 \in M \setminus \{0\}$ and $n_1, n_2 \in N \setminus \{0\}$. Assume that $m_1 \otimes_k n_1 = m_2 \otimes_k n_2$. Prove that there exists $\lambda \in k \setminus \{0\}$ such that $m_2 = \lambda m_1$ and $n_2 = \lambda^{-1} n_1$.
(2) Let X be a subspace of $M \otimes_k N$ comprising only elementary tensors. Prove that

- either there exist $m \in M$ and a subspace X_N of N such that $X = m \otimes_k X_N$,
- or there exist $n \in N$ and a subspace X_M of M such that $X = X_M \otimes_k n$.

Exercise 1.4.9 Let M be a finite dimensional k-vector space. Assume that the inverse image of Id_M by the isomorphism

$$M^* \otimes_k M \xrightarrow{\sim} \text{Hom}(M, M)$$

defined in Proposition 1.2.14 is $\sum_{i \in I} m'_i \otimes_k m_i$ (where $m'_i \in M^*$ and $m_i \in M$).

(1) Prove that for all $m \in M$,

$$\sum_{i \in I} m'_i(m) m_i = m .$$

(2) Assume that $x \in M^* \otimes_k M$ has image $\phi \in \text{Hom}(M, M)$ through the above isomorphim. Prove that

$$x = \sum_{i \in I} m'_i \cdot \phi \otimes_k m_i = \sum_{i \in I} m'_i \otimes_k \phi(m_i) .$$

Exercise 1.4.10 Assume that the characteristic of k is not 2. Let M and N be two k-vector spaces.

(1) Establish an isomorphism $S(M \oplus N) \xrightarrow{\sim} S(M) \otimes S(N)$.
(2) Establish an isomorphism $\Lambda(M \oplus N) \xrightarrow{\sim} \Lambda(M) \otimes \Lambda(N)$.

Chapter 2
On Representations

2.1 Generalities on Representations

2.1.1 Introduction

When groups were discovered, during the XIXth century, there were no "abstract groups" (like *"a set endowed with a law which is associative, etc."*). Groups were given as *permutation groups*, and most of the time as acting on roots of polynomials by preserving the algebraic relations between them.

For example, the group which is now known as the dihedral group of order 8, denoted by D_8, and which is defined by generators σ and τ subject to relations $\sigma^2 = \tau^2 = (\sigma\tau)^4 = 1$, was known as the group which permutes the four complex roots of $X^4 - 2$ by preserving their rational relations.

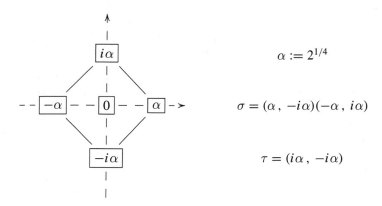

$$\alpha := 2^{1/4}$$

$$\sigma = (\alpha, -i\alpha)(-\alpha, i\alpha)$$

$$\tau = (i\alpha, -i\alpha)$$

One sees that this group may also be seen as the group of isometries of $\mathbb{R}^2$ which preserve a square centered in 0: it is *"represented"* as a subgroup of $\mathrm{GL}_2(\mathbb{R})$. Choosing $(\alpha, i\alpha)$ as a basis of $\mathbb{C}$ (identified with $\mathbb{R}^2$), we see that

© Springer Nature Singapore Pte Ltd. 2017
M. Broué, *On Characters of Finite Groups*, Mathematical Lectures from Peking University, https://doi.org/10.1007/978-981-10-6878-2_2

$$\sigma \mapsto \begin{pmatrix} 0 & -1 \\ -1 & 0 \end{pmatrix} \quad , \quad \tau \mapsto \begin{pmatrix} 1 & 0 \\ 0 & -1 \end{pmatrix} .$$

A representation of a group, roughly, consists in viewing that group as acting on a mathematical object (the set of roots of a polynomial, or a vector space, or a topological space, etc.) preserving the underlying structure.

2.1.2   General Representations

First definitions.
Let C be a category:

- C may be the category **Top** of topological spaces, in which case its objects are topological spaces and its morphisms are the continuous maps between the objects,
- C may be the category **Vect**$_k$ of vector spaces over the field k, in which case its objects are k-vector spaces and its morphisms are the linear maps between the objects, or the subcategory **vect**$_k$ whose objects are the finite dimensional k-vector spaces and morphisms are again the linear maps between the objects (such a subcategory, where the objects run over a subclass, is called a full subcategory),
- C may be the category **Set** of sets, in which case its objects are sets and its morphisms are the maps between the objects, or the full subcategory **set**$_k$ of finite sets,
- etc.

For X and Y objects of C, the set of morphisms from X to Y will be denoted by $\mathrm{Mor}_C(X, Y)$ or simply $\mathrm{Mor}(X, Y)$.

In the case where $C = $ **Vect**$_k$, $\mathrm{Mor}(X, Y)$ is the space $\mathrm{Hom}(X, Y)$ (sometimes denoted, when necessary, $\mathrm{Hom}_k(X, Y)$) of k-linear maps from X to Y.

If X is an object of C, the group $\mathrm{Aut}(X)$ of automorphisms of X consists of all morphisms from X to X which admit an inverse morphism.

- If $C = $ **Top**, $\mathrm{Aut}(X)$ is the group of all homeomorphisms of X into itself,
- if $C = $ **Vect**$_k$, $\mathrm{Aut}(X) = \mathrm{GL}(X)$, the *general linear group* comprising all bijective linear endomorphisms of X,
- if $C = $ **Set**, $\mathrm{Aut}(X) = \mathfrak{S}(X)$, the *symmetric group* of X, group of all bijections of X into itself.

Let now G be a group.

Definitions 2.1.1 • A *representation* of G on C (or a C-*representation* of G) is a pair (X, ρ) where

     ∗ X is an object of C,
     ∗ ρ is a group morphism from G into the group $\mathrm{Aut}(X)$ of automorphisms of X.

- A representation (X, ρ) is said to be *faithful* if the morphism ρ is injective.
- A *morphism* between representations (X, ρ) and (X', ρ') of G on the same category $\mathcal{C}$ is a morphism $f : X \to X'$ in $\mathcal{C}$ such that, for all $g \in G$, the following diagram is commutative :

$$
\begin{array}{ccc}
X & \xrightarrow{\;f\;} & X' \\
{\scriptstyle \rho(g)}\big\downarrow & & \big\downarrow{\scriptstyle \rho'(g)} \\
X & \xrightarrow{\;f\;} & X'
\end{array}
$$

If (X, ρ) is a representation, we say that G *acts on* X.

The reader is invited to define the category $\mathbf{Rep}(G, \mathcal{C})$ of representations of G on $\mathcal{C}$, and in particular to check the following lemma.

Lemma 2.1.2 *A morphism* $f : (X, \rho) \to (X', \rho')$ *in* $\mathbf{Rep}(G, \mathcal{C})$ *is an* isomorphism *if and only if the morphism* $f : X \to X'$ *is an isomorphism in* $\mathcal{C}$.

In particular, two representations (X, ρ) and (X, ρ') on the same object X are isomorphic (one says then that they are *equivalent*) if there exists an automorphism f of X such that, for all $g \in G$,

$$
\rho'(g) = f \cdot \rho(g) \cdot f^{-1} ,
$$

i.e, if ρ and ρ' are "uniformly conjugate".

Notation 2.1.3 If (X, ρ) and (X', ρ') are isomorphic, we write

$$
(X, \rho) \simeq (X', \rho') .
$$

We end this paragraph with a remark on morphisms of representations.
Let (X, ρ) and (Y, σ) be $\mathcal{C}$-representations of G.
The group $\mathrm{Aut}(X) \times \mathrm{Aut}(Y)$ acts on the set $\mathrm{Mor}(X, Y)$ of morphisms from X to Y, by the formula

$$
\text{for } \alpha \in \mathrm{Aut}(X),\ \beta \in \mathrm{Aut}(Y),\ f \in \mathrm{Mor}(X, Y),\ (\alpha, \beta) \cdot f := \beta f \alpha^{-1} .
$$

Composed with the diagonal morphism

$$
G \to \mathrm{Aut}(X) \times \mathrm{Aut}(Y),\ g \mapsto (\rho(g), \sigma(g)),
$$

that operation defines an action of G on $\mathrm{Mor}(X, Y)$:

$$
\text{for } g \in G,\ f \in \mathrm{Mor}(X, Y),\ g \cdot f := \sigma(g) f \rho(g)^{-1} ,
$$

that is, the following diagram is commutative:

$$X \xrightarrow{\ f\ } Y$$

$$\rho(g) \downarrow \qquad \downarrow \sigma(g)$$

$$X \xrightarrow{\ g \cdot f\ } Y$$

The following lemma is then clear.

Lemma 2.1.4 *The set of morphisms* $(X, \rho) \to (Y, \sigma)$ *is the set of fixed points of G in its action on* $\mathrm{Mor}(X, Y)$.

Case where objects have elements.

Remark 2.1.5 Although it is indeed the case for the categories **Top**, **Vect**$_k$, **Set**, the objects of a category need not be "sets containing elements" and morphisms need not "act on elements": for example, the objects of the category **Rep**$(G, \mathcal{C})$ are not "sets containing elements".

Nevertheless, if this is the case, and when there is no ambiguity, if x is an element of X, we shall write

$$gx := \rho(g)(x).$$

In that case, a morphism $(X, \rho) \to (X', \rho')$ is a morphism $f : X \to X'$ in $\mathcal{C}$ such that, for all $g \in G$ and $x \in X$,

$$f(gx) = gf(x).$$

We add a few technical remarks for the case where X contains elements.

If (X, ρ) is a representation of G on $\mathcal{C}$ and if $x \in X$, the fixator of x in G is the subgroup defined by

$$G_x := \{g \in G \mid gx = x\}.$$

The proof of the following lemma is left as an exercise to the reader.

Lemma 2.1.6 *Let* (X, ρ) *be a representation of G.*

(1) For $x \in X$ *and* $g \in G$, *we have*

$$G_{gx} = gG_x g^{-1}.$$

(2) If $f : (X, \rho) \xrightarrow{\sim} (Y, \sigma)$ *is an isomorphism of representations, then for* $x \in X$ *we have* $G_{f(x)} = G_x$.

Action of $\mathrm{Aut}(G)$ *on representations.*

Let us denote by $\mathrm{Aut}(G)$ the group of automorphisms of G.

Any element $g \in G$ defines an element $\mathrm{Inn}(g)$ acting on the left by

$$\mathrm{Inn}(g) : G \xrightarrow{\sim} G , \quad h \mapsto ghg^{-1},$$

and we denote by $\mathrm{Inn}(G)$ the group of all $\mathrm{Inn}(g)$ for $g \in G$.

The group $\mathrm{Inn}(G)$ is *normal* in $\mathrm{Aut}(G)$, since for all $a \in \mathrm{Aut}(G)$ and $g \in G$,

$$a\mathrm{Inn}(g)a^{-1} = \mathrm{Inn}(a(g)).$$

The group $\mathrm{Aut}(G)$ acts on $\mathcal{C}$-representations, as we shall see now.
Let $a \in \mathrm{Aut}(G)$.

- Whenever (X, ρ) is a $\mathcal{C}$-representation of G, we set ${}^a(X, \rho) := (X, {}^a\rho)$ where ${}^a\rho : G \to \mathrm{Aut}(X)$ is the group morphism defined by ${}^a\rho := \rho \cdot a^{-1}$.

It is immediate to check that ${}^{aa'}(X, \rho) = {}^a\left({}^{a'}(X, \rho)\right)$.
- If $f : X \to X'$ induces a morphism of representations

$$(X, \rho) \to (X', \rho'),$$

then it also induces a morphism of representations

$${}^a(X, \rho) \to {}^a(X', \rho').$$

Thus in particular $\mathrm{Aut}(G)$ *acts on the set of isomorphism classes of $\mathcal{C}$-representations.*
- For $g \in G$, $\rho(g)$ is an isomorphism of representations

$$\rho(g) : {}^{\mathrm{Inn}(g)}(X, \rho) \xrightarrow{\sim} (X, \rho).$$

Thus in particular $\mathrm{Inn}(G)$ acts trivially on the set of isomorphism classes of $\mathcal{C}$-representations.

The group $\mathrm{Out}(G) := \mathrm{Aut}(G)/\mathrm{Inn}(G)$ of *outer automorphisms* then acts on the set of isomorphism classes of $\mathcal{C}$-representations.

2.2 Set-Representations

2.2.1 Union and Product

Given sets X and Y, we know how to construct

- their (disjoint) union $X \sqcup Y$,
- their product $X \times Y$,

and if Z is a set, we have

$$Z \times (X \sqcup Y) = (Z \times X) \sqcup (Z \times Y).$$

Now given a group G and representations (on **Set**) (X, ρ) and (Y, σ), we define their union and their product by

$$(\rho \sqcup \sigma)(g)(z) := \begin{cases} \rho(g)(z) & \text{if } z \in X, \\ \sigma(g)(z) & \text{if } z \in Y, \end{cases}$$

$$(\rho \times \sigma)(g)(x, y) := (\rho(g)(x), \sigma(g)(y)) \quad \text{for } x \in X, y \in Y.$$

It is clear that distributivity of the product on the union remains valid for representations.

Let (X, ρ) be a **Set**-representation of G.

If X_1 is a subset of X which is stable by all the bijections $\rho(g)$ for $g \in G$, it gives rise to a *subrepresentation* (X_1, ρ_1) of (X, ρ) (where $\rho_1(g)$ is the restriction of $\rho(g)$ to X_1).

Example 2.2.1 Let (X, ρ) be a representation of G, and let $x \in X$. Then the subset of X defined by $X_1 := \{\rho(g)(x) \mid g \in G\}$ is stable by G, hence defines a subrepresentation.

Let X_1' be the complement of X_1 in X. Then X_1' is also stable by all the $\rho(g)$ for $g \in G$, and we have an isomorphism of representations (with obvious notations)

$$(X, \rho) \simeq (X_1, \rho_1) \sqcup (X_1', \rho_1').$$

2.2.2   Transitive Representations

The proof of the following proposition is immediate, and left to the reader as an exercise.

Proposition – Definition 2.2.2 *Let (X, ρ) be a **Set**-representation of G, $X \neq \emptyset$. The following assertions are equivalent.*

  (i) *(X, ρ) is "simple" or "irreducible", that is, if (X_1, ρ_1) is a subrepresentation of (X, ρ), then either $X_1 = X$ or $X_1 = \emptyset$.*
 (ii) *(X, ρ) is "indecomposable", that is, if $(X, \rho) \simeq (X_1, \rho_1) \sqcup (X_1', \rho_1')$, then either $X_1 = \emptyset$ or $X_1' = \emptyset$.*
(iii) *(X, ρ) is "transitive", that is, for all $x, y \in X$, there exists $g \in G$ such that $y = \rho(g)(x)$.*

Example 2.2.3 Let H be a subgroup of G. We denote by ρ_H the representation (by left multiplication) of G on the set G/H of left cosets of G modulo H. Thus, for $g, x \in G$, we have

$$\rho_H(g)(xH) := gxH .$$

(1) The representation $(G/H, \rho_H)$ is transitive. Indeed, given xH and $yH \in G/H$, we have $yH = \rho_H(yx^{-1})(xH)$.

(2) The fixator of xH is xHx^{-1}. Indeed, it can be proved either by a direct computation or by applying Lemma 2.1.6, (1).

We shall see (Theorem 2.2.5 below) that the above example is "universal".

Let us first check that any representation is isomorphic to a disjoint union of transitive representations.

Let (X, ρ) be a **Set**-representation of G.

We define the equivalence relation $\sim_G$ on X as follows:

$$(x \sim_G y) \iff (\exists g \in G)(y = \rho(g)(x)) .$$

We denote by $G\backslash X$ the set of equivalence classes of the above relation.

The proof of the following proposition is obvious.

Proposition 2.2.4 *(1) Any equivalence class Ω of $\sim_G$ defines a transitive subrepresentation (Ω, ρ_Ω) of (X, ρ).*

(2) Thus

$$(X, \rho) \simeq \bigsqcup_{\Omega \in G\backslash X} (\Omega, \rho_\Omega) .$$

2.2.3 Classification of Transitive Representations

Let us now classify all transitive representations. The next theorem shows that the isomorphism classes of transitive representations of G are naturally parametrized by conjugacy classes of subgroups of G.

Theorem 2.2.5 *(1) Let (X, ρ) be a transitive representation of G. For $x \in X$, let G_x denote the fixator of x in G. Then the map*

$$G/G_x \to X , \quad gG_x \mapsto \rho(g)(x)$$

defines an isomorphism $(G/G_x, \rho_{G_x}) \overset{\sim}{\longrightarrow} (X, \rho)$:

$$
\begin{array}{ccc}
G/G_x & \overset{\sim}{\longrightarrow} & X \\
{\scriptstyle g}\downarrow & & \downarrow{\scriptstyle \rho(g)} \\
G/G_x & \overset{\sim}{\longrightarrow} & X
\end{array}
$$

*(2) For subgroups H and H' of G, the representations $(G/H, \rho_H)$ and $(G/H', \rho_{H'})$
      are isomorphic if and only if H and H' are conjugate.*

Proof The proof of (1) is easy and left to the reader.

(2) Assume first that H and H' are conjugate. Since $(G/H, \rho_H)$ is transitive and
since H is the fixator of $H \in G/H$ (yes!), it follows from Lemma 2.1.6, (1), that
H' is the fixator of an element of G/H. Thus by (1) it follows that $(G/H, \rho_H)$ and
$(G/H', \rho_{H'})$ are isomorphic.

Assume now that $(G/H, \rho_H)$ and $(G/H', \rho_{H'})$ are isomorphic. It follows from
Lemma 2.1.6, (2), that both H and H' are fixators of elements of G/H, hence are
conjugate by Lemma 2.1.6, (1).   □

If (X, ρ) is a **Set**-representation, the cardinality of X is called the *degree* of the
representation.

Corollary 2.2.6 *If G is a finite group, the degree of any transitive representation of
G divides the order of G.*

Notation 2.2.7 From now on, we also call G-*set* a set X endowed with an action of
G, that is, a group morphism $\rho : G \to \mathfrak{S}(X)$.

The category **Rep**$(G, \mathbf{Set})$ is also denoted $_G\mathbf{Set}$.

Notation 2.2.8 When a finite group G acts on an object X "with elements" (e.g. a
set, a vector space, or a set of morphisms $\mathrm{Mor}(Y, Z)$), we denote by $\mathrm{Fix}^G(X)$ the
subset of fixed points of X under G.

2.2.4   Burnside's Marks

From now on, all our G-sets are assumed to be finite. In other words, we consider
the category $_G\mathbf{set}$.

Definition and properties of marks.

Definition 2.2.9 Let G be a finite group and let X be a (finite) G-set. *The mark of
X is the function m_X defined on the set of all subgroups of G, such that, for any
subgroup H of G,*

$$m_X(H) := |\mathrm{Fix}^H(X)| .$$

Let us denote by $\mathcal{S}(G)$ the set of *conjugacy classes of subgroups* of G.
Define the following partial order on $\mathcal{S}(G)$:

$$(C \leq C') :\Longleftrightarrow (\exists H \in C)(\exists H' \in C')(H \subseteq H') .$$

Notice that:

- the value $m_X(H)$ depends only on the conjugacy class of the group H,
- by Theorem 2.2.5, the set of isomorphism classes of transitive G-sets is in natural bijection with the set $\mathcal{S}(G)$ of conjugacy classes of subgroups of G.

Thus the function mark induces a function on $\mathcal{S}(G) \times \mathcal{S}(G)$:

$$m(C, C') := m_{G/H'}(H) \quad \text{for } H \in C \text{ and } H' \in C'.$$

For H a subgroup of G, we denote by

$$N_G(H) := \{g \in G \mid gHg^{-1} = H\}$$

the *normalizer* of H in G.

Lemma 2.2.10 *(1)* $m(C, C') \neq 0$ *if and only if* $C \leq C'$.
(2) $m(C, C) = |N_G(H)/H|$ *for* $H \in C$.

Proof (1) A group H fixes an element in G/H' if and only if it is contained in the fixator of an element of G/H'. But the fixators of elements in G/H' are the conjugates of H'. Hence $m_{G/H'}(H) \neq 0$ if and only if H is contained in H' up to conjugation.

(2) We have

$$\begin{aligned}
\mathrm{Fix}^H(G/H) &= \{gH \in G/H \mid HgH = gH\} \\
&= \{gH \in G/H \mid g^{-1}HgH = H\} \\
&= \{gH \in G/H \mid g \in N_G(H)\} = N_G(H)/H.
\end{aligned}$$

$\square$

Marks characterize representations.

Proposition 2.2.11 *Let X and Y be finite G-sets. The following assertions are equivalent.*

(i) X and Y are isomorphic as G-sets.
(ii) $m_X = m_Y$.

The proof will use the notion of *Möbius function of a poset* — the reader may refer to [Aig79, Chap. IV] for details and examples. It may be viewed as an easy computation of the inverse of a non singular "triangular matrix on a partially ordered set".

Given the function $m : \mathcal{S}(G) \times \mathcal{S}(G) \to \mathbb{Q}$, let us define the function $\mu : \mathcal{S}(G) \times \mathcal{S}(G) \to \mathbb{Q}$ by the following conditions:

(1) for all $C, D \in \mathcal{S}(G)$, $\mu(C, D) \neq 0$ implies $C \leq D$,
(2) for all $C \in \mathcal{S}(G)$, $\mu(C, C)m(C, C) = 1$,

(3) if $C < E$, then

$$\sum_{C \leq D \leq E} \mu(C, D) m(D, E) = 0.$$

Note that condition (3) allows to compute $\mu(C, E)$ knowing $\mu(C, D)$ for $C \leq D < E$, hence defines $\mu(C, E)$ for all $C \leq E$ by induction since $m(E, E) \neq 0$. Note also that μ takes indeed its values in $\mathbb{Q}$.

The following lemma is an inversion formula.

Lemma 2.2.12 *Let $f : S(G) \to \mathbb{Q}$ be a function. Define $\widehat{f} : S(G) \to \mathbb{Q}$ by*

$$\widehat{f}(C) := \sum_{D \geq C} m(C, D) f(D).$$

Then

$$f(C) = \sum_{D \geq C} \mu(C, D) \widehat{f}(D).$$

Proof It is an immediate computation:

$$\sum_{D \geq C} \mu(C, D) \widehat{f}(D) = \sum_{D \geq C} \mu(C, D) \Big(\sum_{E \geq D} m(D, E) f(E) \Big)$$

$$= \sum_{E \geq C} f(E) \Big(\sum_{C \leq D \leq E} \mu(C, D) m(D, E) \Big)$$

$$= f(C).$$

$\square$

Proof of Proposition 2.2.11 For $D \in S(G)$ (a conjugacy class of subgroups of G), let us denote by X_D the transitive G-set associated with some $H \in D$.

Let X be a G-set, which is then isomorphic to a disjoint union of some X_D for $D \in S(G)$. We denote by $f_X(D)$ the multiplicity of G-sets isomorphic to X_D in such a decomposition, which we write

$$X \simeq \bigsqcup_{D \in S(G)} f_X(D).X_D.   \qquad (*)$$

To prove Proposition 2.2.11, it suffices to prove that the function m_X determines the function f_X.

By equality (*) above, since by definition $m(C, D) = m_{X_D}(C)$ and since $m(C, D) \neq 0 \Rightarrow C \leq D$, we have

$$m_X(C) = \sum_{D \geq C} f_X(D) m(C, D).$$

Lemma 2.2.12 shows then that

$$f_X(C) = \sum_{D \geq C} \mu(C, D) m_X(D),$$

which proves that m_X determines f_X.   □

2.3 Linear Representations

Throughout this section, k is a field. We shall only consider representations of finite groups on finite dimensional vector spaces – in other words, vect$_k$-representations.

2.3.1 Generalities

A (finite dimensional) *k-linear representation of G* is a pair (M, ρ) where M is a finite dimensional k-vector space and $\rho : G \to GL(M)$ is a group morphism. We shall often abbreviate (if there is no ambiguity) (M, ρ) by M.

The set of morphisms from (M, ρ) to (N, σ) is the set of fixed points of G acting on the space $\mathrm{Hom}_k(M, N)$ of linear maps from M to N (see Lemma 2.1.4). This set of morphisms will be denoted by $\mathrm{Hom}_{kG}(M, N)$.

Thus the category $\mathbf{Rep}(G, \mathbf{vect}_k)$ may be seen as follows:

- its objects are the pairs (M, ρ), where M is a finite dimensional k-vector space and $\rho : G \to GL(M)$ a group morphism – or, simpler, the objects are *finite dimensional k-vector spaces endowed with a k-linear action of G*,
- the set of morphisms between objects M and N is

$$\mathrm{Hom}_{kG}(M, N) = \left\{ \alpha \in \mathrm{Hom}_k(M, N) \mid (\forall g \in G)(\alpha g = g\alpha) \right\}.$$

The dimension of M is called the *degree* of the representation.

The *trivial representation* is $(k, \rho_{\mathrm{triv}})$ where $\rho_{\mathrm{triv}} : G \to GL(k) = k^{\times}$ is the trivial morphism (i.e. $\rho_{\mathrm{triv}}(g) = 1$ for all $g \in G$).

Matrix representations.

Assume M has dimension r. The choice of a basis of M defines a group isomorphism $GL(M) \xrightarrow{\sim} GL_r(k)$ (where $GL_r(k)$ denotes the group of all invertible $r \times r$ matrices with entries in k).

Thus a representation of G may be seen as a pair (r, ρ) where r is a natural integer and $\rho : G \to GL_r(k)$ is a group morphism.

In this context, two representations (r, ρ) and (s, σ) are isomorphic (equivalent) if and only if $r = s$ and there is an element $f \in GL_r(k)$ such that

$$\forall g \in G, \ \sigma(g) = f \cdot \rho(g) \cdot f^{-1},$$

i.e. if the matrices $\rho(g)$ and $\sigma(g)$ (for $g \in G$) are *uniformly similar*.

Extension of scalars.

Assume that k is a subfield of a field k'.

• Given a k-representation (M, ρ), its extension of scalars is the k'-representation $(k' \otimes_k M, k' \otimes_k \rho)$, where we denote by $k' \otimes_k \rho$ the morphism $G \to \mathrm{GL}(k' \otimes_k M)$ defined by

$$g \mapsto \mathrm{Id}_{k'} \otimes \rho(g).$$

A basis of M over k is identified with a basis of $k' \otimes_k M$ over k'. Thus, from the point of view of a matrix representation $\rho : G \to \mathrm{GL}_r(k)$, extending the scalars consists in viewing the matrix entries of $\rho(g)$ as elements of k'.

• Conversely, a k'-representation (M', ρ') is said to be *rational over k* if there exists a k-representation (M, ρ) such that

$$(M', \rho') \simeq (k' \otimes_k M, k' \otimes_k \rho).$$

From the point of view of matrix representations, a representation $G \to \mathrm{GL}_r(k')$ is rational over k if there exists an element $U \in \mathrm{GL}_r(k')$ such that, for all $g \in G$, all the entries of $U\rho(g)U^{-1}$ belong to k.

Construction of representations.

If (M, ρ) and (N, σ) are representations of G, one may build

• their direct sum $(M \oplus N, \rho \oplus \sigma)$ (the construction is left to the reader), the analog of the disjoint union of **Set**-representations,
• their tensor product $(M \otimes_k N, \rho \otimes_k \sigma)$ (the construction is left to the reader), the analog of the product of **Set**-representations.

If M_1 is a subspace of M which is stable by the action of all $\rho(g)$ for $g \in G$, we get a *subrepresentation* (M_1, ρ_1) of (M, ρ) (where $\rho_1(g)$ is the restriction of $\rho(g)$ to M_1).

In this case, we also have a *quotient representation* $(M/M_1, \overline{\rho})$ (where $\overline{\rho}(g)$ is the automorphism of M/M_1 induced by $\rho(g)$).

  ○! Contrary to what happens for **Set**-representations, given a representation (M, ρ) and a G-stable subspace M_1, such a subspace need not have a G-stable complement (see the following exercise).

Exercise 2.3.1 Assume that G is the cyclic (additive) group $\mathbb{Z}/p\mathbb{Z}$ where p is a prime number, and let k be a field of characteristic p.

Consider the morphism

$$G \to GL_2(k) , \quad \lambda \mapsto \begin{pmatrix} 1 & \lambda \\ 0 & 1 \end{pmatrix} .$$

Prove that the line generated by the first standard basis vector of k^2 is G-stable but has no G-stable complement.

Indecomposable and irreducible representations.
As for **Set**-representations, we say that a representation (M, ρ) is

- *simple* (or *irreducible*) if $M \neq \{0\}$ and the only G-stable subspaces are M and $\{0\}$,
- *indecomposable* if $M \neq \{0\}$ and $(M, \rho) \simeq (M_1 \oplus M_1', \rho_1 \oplus \rho_1')$ imply either $M_1 = \{0\}$, or $M_1' = \{0\}$.

① Contrary to what happens for **Set**-representations, a representation may be indecomposable without being irreducible (see below Exercise 2.3.2).

① Nevertheless, we shall see below (3.1.5) that over *characteristic zero* fields, again the indecomposable and the irreducible representations coincide (for a *finite* group).

Exercise 2.3.2 Prove that the representation defined in Exercise 2.3.1 is not irreducible but is indecomposable.

The first assertion of the following theorem is a consequence of an easy induction on the degree of the representation. The second assertion is a much deeper result, known as *Krull–Schmidt Theorem*, which won't be proved here.

Theorem 2.3.3 *Let (M, ρ) be a representation of finite degree.*

(1) It is a direct sum of indecomposable representations.
(2) If G is finite, and if $(M, \rho) \simeq \bigoplus_{i \in I} (M_i, \rho_i) \simeq \bigoplus_{j \in J} (M_j', \rho_j')$ where the (M_i, ρ_i) and the (M_j', ρ_j') are indecomposable, there is a bijection $\alpha : I \xrightarrow{\sim} J$ such that, for all $i \in I$, $(M_i, \rho_i) \simeq (M_{\alpha(i)}', \rho_{\alpha(i)}')$.

From set-representations to vect$_k$-representations.
Let (Ω, ρ) be a **set**-representation of G. One denotes by $k\Omega$ the k-vector space with basis Ω. Then it is clear that ρ induces a group morphism

$$k\rho : G \to GL(k\Omega) .$$

The reader may check that this construction transforms

- the disjoint union of **set**-representations into the direct sum of the corresponding **vect**$_k$-representations,
- the product of **set**-representations into the tensor product of the corresponding **vect**$_k$-representations.

Remark 2.3.4 Using the notion of *functor* between two categories (see Appendix E), one can check easily that the above construction induces a functor

$$\mathbf{Rep}(G, \mathbf{set}) \to \mathbf{Rep}(G, \mathbf{vect}_k).$$

ⓘ **Attention** ⓘ Even if (Ω, ρ) is transitive with degree at least 2, $(k\Omega, k\rho)$ is *not* irreducible.

Indeed, the element $S\Omega := \sum_{\omega \in \Omega} \omega \in k\Omega$ spans a line which is stable under G.

Definition 2.3.5 The group G acts on itself by left multiplication. The corresponding k-linear representation is called the *regular representation* of G.

The preceding example is used in the following proposition.

Proposition 2.3.6 *Assume that k has nonzero characteristic p which divides the order $|G|$ of G. Then the line kSG has no G-stable complement in the regular representation kG.*

Proof Let us choose a complement of kSG in kG, that is, let us choose a projector $\pi : kG \twoheadrightarrow kSG$. Hence $\pi(SG) = SG$.

Now if the corresponding complement were G-stable, we would have $g\pi(x) = \pi(gx)$ for all $x \in kG$, which would imply $\pi(SG) = |G|\pi(1) = 0$ since the characteristic of k divides $|G|$. This is a contradiction. $\square$

ⓘ Nevertheless, we shall see below (3.1.3) that over characteristic zero fields such a situation cannot happen, since then for every subrepresentation there is a G-stable complement.

Contragredient, homomorphisms and tensor products.

Let (M, ρ) and (N, σ) be $\mathbf{vect}_k$-representations of G, which we denote (by abuse of notation) by their "vector spaces parts" M and N.

• There is a natural representation associated with $\mathrm{Hom}_k(M, N)$ (the space of linear maps from M to N), defined by the following formula:

$$\text{for } \alpha \in \mathrm{Hom}_k(M, N) \text{ and } g \in G, \ g \cdot \alpha := \sigma(g)\alpha\rho(g)^{-1}.$$

• In particular, there is a natural representation associated with the dual $M^* := \mathrm{Hom}_k(M, k)$ of linear maps from M to k, defined by the following formula:

$$\text{for } \alpha \in \mathrm{Hom}_k(M, k) \text{ and } g \in G, \ g \cdot \alpha := \alpha\rho(g)^{-1}.$$

This representation is called the *contragredient representation* of (M, ρ).

Exercise 2.3.7 Compare the matrices associated with a representation and the matrices associated with the contragredient representation with respect to a pair of dual bases.

• There is a natural representation associated with the tensor space $M \otimes_k N$, defined by the following formula:

for $m \in M$, $n \in N$ and $g \in G$, $g \cdot (m \otimes_k n) := \rho(g)(m) \otimes_k \sigma(g)(n)$.

Proposition 2.3.8 *(1) The morphism*

$$M^* \otimes_k N \to \mathrm{Hom}_k(M, N) , \; \varphi \otimes_k n \mapsto (m \mapsto \varphi(m)n) ,$$

defines a morphism of representations.
(2) If M and N are finite dimensional, this is an isomorphism.

Proof (1) Applying $g \in G$ to
• the elementary tensor $\varphi \otimes_k n$ gives $\varphi\rho(g)^{-1} \otimes_k \sigma(g)(n)$,
• the map $(m \mapsto \varphi(m)n)$ gives the map $m \mapsto \varphi(\rho(g)^{-1}(m))\sigma(g)(n)$, and this establishes (1).
(2) follows from the fact that the above morphism is an isomorphism if M and N are finite dimensional (see Proposition 1.2.14).   □

Issai Schur (1875–1941)

Irreducible representations, Schur's Lemma.
  We recall that a representation (S, ρ) (S a k-vector space, $\rho : G \to \mathrm{GL}(S)$) is said to be *irreducible* if $S \neq \{0\}$ and the only G-stable subspaces of S are $\{0\}$ and S.

Proposition 2.3.9 (Schur's Lemma) *Let (S, ρ) and (S', ρ') be irreducible representations of G.*

(1) If (S, ρ) and (S', ρ') are not isomorphic, then $\mathrm{Hom}_{kG}(S, S') = \{0\}$.
(2) $\mathrm{End}_{kG}(S) := \mathrm{Hom}_{kG}(S, S)$ is a division k-algebra, a finite extension of k.
(3) If k is algebraically closed, $\mathrm{End}_{kG}(S) = k\mathrm{Id}_S$.

Proof Let $f \in \text{Hom}_{kG}(S, S')$. Assume $f \neq 0$. Then

- $\ker(f)$ is a proper G-stable subspace of S, hence $\ker(f) = 0$;
- $\text{im}(f)$ is a nonzero subspace of S', hence $\text{im}(f) = S'$.

This proves that a nonzero kG-morphism from S to S' is an isomorphism, whence (1) and (2).

To prove (3), we may for example notice that any $f \in \text{End}_{kG}(S)$ has an eigenvalue λ. Then $\ker(f - \lambda \text{Id}_S)$ is a G-stable nonzero subspace of S, hence $\ker(f - \lambda \text{Id}_S) = S$ and $f = \lambda \text{Id}_S$.   □

Representations of degree 1.

A *representation of degree 1* is obviously irreducible. It corresponds to a morphism

$$G \rightarrow \text{GL}_1(k) = k^\times .$$

Since $k^\times$ is abelian, morphisms $G \rightarrow k^\times$ correspond naturally to morphisms

$$G/[G, G] \rightarrow k^\times ,$$

where $[G, G]$ denotes the *derived subgroup of G*, which is the subgroup generated by the *commutators* $[g, h] := ghg^{-1}h^{-1}$ for $g, h \in G$.

Example 2.3.10 Consider the dihedral group $G = \langle \sigma, \tau \rangle$ of order 8 mentioned in the introduction of this chapter.

Its derived subgroup is equal to its center, the subgroup of order 2 generated by $(\sigma\tau)^2$. It is not difficult to check that $G/[G, G]$ is isomorphic to the direct product of two groups of order 2.

Hence $G/[G, G]$ has four degree 1 representations over a field of characteristic different from 2, namely

$$\rho_1 : \sigma \mapsto 1, \tau \mapsto 1; \quad \rho_\sigma : \sigma \mapsto -1, \tau \mapsto 1;$$
$$\rho_\tau : \sigma \mapsto 1, \tau \mapsto -1; \rho_{\sigma\tau} : \sigma \mapsto -1, \tau \mapsto -1.$$

If (M, ρ) is any representation and if (k, σ) is a degree one representation, the tensor product $(M \otimes_k k, \rho \otimes_k \sigma)$ is naturally equivalent to the representation $(M, \rho\sigma)$ where

$$\rho\sigma : G \rightarrow \text{GL}(M) , \quad g \mapsto \sigma(g)\rho(g) .$$

The following lemma is immediate.

Lemma 2.3.11 *Let (S, ρ) be an irreducible representation of G and let (k, σ) be a degree one representation. Then the tensor product $(S, \sigma\rho)$ is still irreducible.*

⚠ **Attention** ⚠ Tensoring an irreducible representation by a nontrivial representation of degree one need not change the equivalence class of the representation.

As an exercise, the reader may check that for the dihedral group of order 8, multiplying its degree 2 real representation (as defined in the introduction of this chapter) by any degree 1 representation provides equivalent representations — a fact which will become obvious with characters.

2.3.2 Finite Groups: The Group Algebra

The group algebra.
Let G be a finite group. The *group algebra* kG is the k-vector space kG endowed with multiplication defined by the multiplication of elements of G.

Thus, for $x := \sum_{s \in G} x_s s$ and $y := \sum_{t \in G} y_t t$ in kG, we have

$$xy = \sum_{s,t \in G} x_s y_t st = \sum_{g \in G} \Big(\sum_{s \in G} x_s y_{s^{-1}g} \Big) g \,.$$

Remark 2.3.12 One sees that the group algebra kG may be seen as the convolution algebra of k-valued functions on G.

Let $\tau : kG \to k$ be the linear form on kG defined by

$$\tau(g) = 0 \text{ if } g \neq 1 \text{ and } \tau(1) = 1 \,.$$

The next lemma shows that τ is a *symmetrizing form* on kG (thus giving kG the structure of a *symmetric algebra*).

Lemma 2.3.13 *(1)* τ *is a "class function": for all* $x, y \in kG$, $\tau(xy) = \tau(yx)$ *.*
(2) The map

$$\widehat{\tau} : \begin{cases} kG \to (kG)^* = \mathrm{Hom}_k(kG, k) \\ x \mapsto \big(y \mapsto \tau(xy)\big) \end{cases}$$

is a k-linear isomorphism.

Proof (1) It is clear that, for all $g, h \in G$, $\tau(gh) = \tau(hg)$. This implies (1) by linearity.

(2) The basis $(\widehat{\tau}(g^{-1}))_{g \in G}$ is the dual basis of the basis $(g)_{g \in G}$ of kG. Thus the image of $\widehat{\tau}$ contains a basis of $\mathrm{Hom}_k(kG, k)$, which shows that $\widehat{\tau}$ is an isomorphism. □

We identify functions $G \to k$ with linear forms on kG, and we denote by $\mathrm{F}(G, k)$ the k-vector space of functions on G.

The following property is easy to check.

2.3.14. *For $f \in \mathrm{F}(G, k)$, the inverse image of f under the isomorphism given in assertion (2) of Lemma 2.3.13 is*

$$f^\circ := \sum_{g \in G} f(g^{-1})g \,.$$

Convention 2.3.15 *Let* $f : G \to k$ *be a* class function *on* G, *that is, such that* $f(gh) = f(hg)$ *for all* $g, h \in G$. *We identify* f *(without changing notation) with the corresponding linear form* $f : kG \to k$, *which is also a class function, i.e.* $f(xy) = f(yx)$ *for all* $x, y \in kG$. *Let* $\mathrm{CF}(G, k)$ *denote the space of class functions on* G *(or* kG*).*

The following lemma will be useful later (see Theorem 3.3.5).

Lemma 2.3.16 *Let* $a \in kG$ *and let* $f \in \mathrm{F}(G, k)$. *Let us denote by* $a \cdot f$ *the element of* $\mathrm{F}(G, k)$ *defined by*

$$(a \cdot f)(x) := f(xa) \text{ for all } x \in kG \,.$$

Then

$$(a \cdot f)^0 = af^0 \,.$$

Proof By definition of f^0, we have $f(x) = \tau(f^0 x)$, hence $(a \cdot f)(x) = f(xa) = \tau(f^0 xa) = \tau(af^0 x)$, which establishes the lemma. □

Let $\mathrm{Cl}(G)$ denote the set of conjugacy classes of G, and for $C \in \mathrm{Cl}(G)$ let us set

* γ_C the *characteristic function* of C, such that $\gamma_C(g) = 0$ if $g \notin C$ and $\gamma_C(g) = 1$ if $g \in C$,
* $SC := \sum_{g \in C} g \in kG$.

For f a class function, we have

$$f = \sum_{C \in \mathrm{Cl}(G)} f(g_C)\gamma_C \text{ and } f^\circ = \sum_{C \in \mathrm{Cl}(G)} f(g_C^{-1})SC \,,$$

where g_C denotes an element of C.

Lemma 2.3.17 *Let* ZkG *denote the center of the group algebra* kG.

(1) The family $(\gamma_C)_{C \in \mathrm{Cl}(G)}$ *is a basis of* $\mathrm{CF}(G, k)$, *and the family* $(SC)_{C \in \mathrm{Cl}(G)}$ *is a basis of* ZkG. *Hence*

$$[\mathrm{CF}(G, k) : k] = |\mathrm{Cl}(G)| \,.$$

(2) The isomorphism of Lemma 2.3.13, (2) restricts to a k-*linear isomorphism* $\mathrm{CF}(G, k) \xrightarrow{\sim} ZkG$. *Hence*

$$[ZkG : k] = |\mathrm{Cl}(G)| \,.$$

kG-*modules and representations of* G.
• Let (M, ρ) be a representation of G. Then the morphism $\rho : G \to GL(M)$ extends by linearity to an *algebra morphism* $\rho : kG \to \mathrm{End}_k(M)$.

Reciprocally, given an algebra morphism $\rho : kG \to \mathrm{End}_k(M)$ (by which we mean in particular that it sends 1 to 1), it restricts to a group morphism $\rho : G \to GL(M)$.
• Now, given an algebra morphism $\rho : kG \to \mathrm{End}_k(M)$ amounts to defining a multiplication

$$kG \times M \to M , \quad (x, m) \mapsto xm$$

(defined by $xm := \rho(x)(m)$) satisfying the conditions (for $x, y \in kG$, $m, n \in M$, $\lambda \in k$)

$$\begin{cases} (x + y)m = xm + ym , & x(ym) = (xy)m \text{ and } 1m = m , \\ x(m + n) = xm + xn , & \lambda(xm) = (\lambda x)m = x(\lambda m) , \end{cases} \quad \text{(mod)}$$

"as if M were a kG-vector space" (except that kG is not a field...). We then say that M is a kG-*module*.

Reciprocally, given a multiplication $kG \times M \to M$ satisfying (mod) defines an algebra morphism $kG \to \mathrm{End}_k(M)$, hence a representation of G on M.

Notation 2.3.18 From now on, we shall speak interchangeably, either of representations, or of kG-modules.

We also call kG-*module* a k-vector space M endowed with an action of G, that is, a group morphism $\rho : G \to GL(M)$.

The category **Rep**$(G, \mathbf{vect}_k)$ is also denoted $_{kG}\mathbf{mod}$.

Notice that a morphism of representations

$$f : (M, \rho) \to (N, \sigma)$$

corresponds to a morphism of kG-modules, i.e. a k-linear map

$$f : M \to N \quad \text{such that } f(xm) = xf(m)$$

for all $x \in kG$ and $m \in M$.

Notice that the regular representation of G, viewed in terms of kG-modules, consists just of the group algebra acted on by itself by left multiplication.

Lemma 2.3.19 *Let* $\rho_{kG} : kG \to \mathrm{End}_k(kG)$ *denote the morphism associated with the regular representation. Then* ρ_{kG} *is injective, i.e. the regular representation of* kG *is faithful.*

Proof Indeed, notice that for all $x \in kG$, $x = \rho_{kG}(x)(1)$. Hence $\rho_{kG}(x) = 0$ implies $x = 0$. □

More Exercises on Chap. 2

Exercise 2.3.20 Let $\mathfrak{S}_n$ denote the n-th symmetric group, the group of all permutations of $\{1, \ldots, n\}$, of order $n!$.

(1) Check that the set $D_4 := \{1, (1\,2)(3\,4), (1\,3)(2\,4), (1\,4)(2\,3)\}$ is a normal subgroup of $\mathfrak{S}_4$.

Deduce that this defines a group morphism $\mathfrak{S}_4 \to \mathfrak{S}_3$.
(2) Prove that this morphism is surjective.

Exercise 2.3.21 (1) Let G be a finite group acting on a finite set Ω. Let $\mathrm{Fix}^G(\Omega)$ denote the set of fixed points of Ω under G, and let $\Omega^{\#}$ denote the set of orbits of G on $\Omega \setminus \mathrm{Fix}^G(\Omega)$. Prove that

$$|\Omega| = |\mathrm{Fix}^G(\Omega)| + \sum_{\omega \in \Omega^{\#}} |\omega| .$$

From now on, we assume that p is a prime number and that G is a nontrivial p-group.
(2) Assume now that $|\Omega|$ is a power of p different from 1. Prove that $|\mathrm{Fix}^G(\Omega)|$ is divisible by p.
(3) Let k be a (commutative) field of characteristic p. Let M be a kG-module. Prove that $\mathrm{Fix}^G(M) \neq 0$.
HINT.

(a) Let $(e_i)_{i=1,\ldots,r}$ be a k-basis of M. Prove that the abelian group generated by $(ge_i)_{(i=1,\ldots,r)(g \in G)}$ is an $\mathbb{F}_p G$-module.
(b) Apply question (2) above to conclude.

(4) Prove that, up to isomorphism, the trivial representation of G is the unique irreducible kG-module.

Exercise 2.3.22 Let G be the dihedral group of order 8, generated by elements σ and τ subject to relations $\sigma^2 = \tau^2 = (\sigma\tau)^4 = 1$.

(1) Check that the map

$$\rho : \begin{cases} G & \to \mathrm{GL}_2(\mathbb{C}) \\ \sigma & \mapsto \begin{pmatrix} 0 & -i \\ i & 0 \end{pmatrix} , \quad \tau \mapsto \begin{pmatrix} 0 & -1 \\ -1 & 0 \end{pmatrix} \end{cases}$$

defines a complex representation of G.
(2) Prove that this representation is rational over $\mathbb{Q}$.

Exercise 2.3.23 Let G be a finite group and let (M, ρ) be a faithful $\mathbb{C}$-linear representation of G.
Assume (M, ρ) irreducible. Prove that the center of G is cyclic.

Exercise 2.3.24 Assume that the (commutative) field k has characteristic $p > 1$. Let $G = \langle g \rangle$ be a cyclic group of order n. We set $n = p^a m$ with $p^a := n_p$ (the largest power of p which divides n) and $m := n_{p'}$ (the largest divisor of n which is prime to p). We assume the field k contains ζ, a primitive m-th root of unity.

(1) For M a kG-module, let $\rho_M : kG \to \mathrm{End}_k(M)$ denote the "structural morphism".

    (a) What are the possible eigenvalues of $\rho_M(g)$?

    (b) Prove that there exists a direct sum decomposition

$$M = M_1 \oplus \cdots \oplus M_m ,$$

    such that the endomorphism $\rho_{M_i}(g) - \zeta^i \mathrm{Id}_{M_i}$ is nilpotent (for all $i = 1, \ldots, m$).

(2) For all pairs of integers (i, j) such that $0 \le i < m$ and $1 \le j \le p^a$, we denote by $M_{i,j}$ the kG-module whose underlying k-vector space is k^j, and where (on the canonical basis of k^j)

$$\rho_{M_{i,j}}(g) := \begin{pmatrix} \zeta^i & 1 & 0 & \cdots & 0 \\ 0 & \zeta^i & 1 & \cdots & 0 \\ \vdots & \vdots & \ddots & \ddots & \vdots \\ 0 & 0 & 0 & \cdots & 1 \\ 0 & 0 & 0 & \cdots & \zeta^i \end{pmatrix}$$

    Prove that, for all (i, j), the kG-module $M_{i,j}$ is indecomposable.

(3) Prove that the family $(M_{i,j})_{0 \le i < m,\, 1 \le j \le p^a}$ is a complete set of representatives of indecomposable kG-modules.

Chapter 3
Characteristic 0 Representations

From now on we assume that

- G is a finite group,
- K is a field of characteristic 0, so $\mathbb{Z}$ is naturally embedded into K.
- All considered KG-modules are finite dimensional.

Ferdinand Georg Frobenius (1849–1917)

3.1 Preliminary: $\frac{1}{|G|} \sum_{g \in G} g$

3.1.1 Fixed Points

The first proposition is a key fact.

© Springer Nature Singapore Pte Ltd. 2017
M. Broué, *On Characters of Finite Groups*, Mathematical Lectures from Peking
University, https://doi.org/10.1007/978-981-10-6878-2_3

For (M, ρ) a K-linear representation of G, let $\mathrm{Fix}^G(M)$ denote the subspace of fixed points of M under the action of G.

Lemma 3.1.1 (1) *Let (M, ρ) be a K-linear representation of G. Then the endomorphism of M defined by*

$$\pi_M^G := \frac{1}{|G|} \sum_{g \in G} \rho(g)$$

is a G-stable projector $M \twoheadrightarrow \mathrm{Fix}^G(M)$.
(2) $\mathrm{Fix}^G(M)$ *has a G-stable complement in M.*
(3) $[\mathrm{Fix}^G(M) : K] = \dfrac{1}{|G|} \sum_{g \in G} \mathrm{tr}\rho(g)$.

Proof (1) It is immediate to check that $(\pi_M^G)^2 = \pi_M^G$, hence that π_M^G is a projector. Its image is obviously contained in $\mathrm{Fix}^G(M)$. Finally, if $x \in \mathrm{Fix}^G(M)$, it is clear that $x = \pi_M^G(x)$.
(2) Since π_M^G commutes with the operation of G, its kernel is stable under G. That kernel is a complement of its image $\mathrm{Fix}^G(M)$.
(3) The dimension of the image of a projector is equal to its trace. That proves (3). $\square$

Corollary 3.1.2 (1) *Let M and N be representations of G. Then*

$$\mathrm{Hom}_{KG}(M, N) = \left\{ \frac{1}{|G|} \sum_{g \in G} g\alpha g^{-1} \mid \alpha \in \mathrm{Hom}_K(M, N) \right\}.$$

(2) *Assume that S and S' are nonisomorphic irreducible G-representations. Then for all $\alpha \in \mathrm{Hom}_K(S, S')$, we have*

$$\frac{1}{|G|} \sum_{g \in G} g\alpha g^{-1} = 0 .$$

Proof (1) By Lemma 2.1.4 and item (1) of the Lemma 3.1.1,

$$\mathrm{Hom}_{KG}(M, N) = \mathrm{Fix}^G(\mathrm{Hom}_K(M, N)) .$$

Now (1) follows from the fact that the action of G on $\mathrm{Hom}_K(M, N)$ is given by $g \cdot \alpha = g\alpha g^{-1}$.
(2) follows from Schur Lemma 2.3.9, (1).

$\square$

3.1.2   Maschke's Theorem

The following theorem expresses the fact that any KG-module M is *semisimple*.

Theorem 3.1.3 (Maschke) *Let M be a K-vector space endowed with an action of G (a KG-module), and let M_1 be a subspace stable under G. Then there is a complement of M_1 which is stable under G.*

Proof Let us denote by $\pi : M \twoheadrightarrow M_1$ a projector onto M_1. Let us consider the G-stable endomorphism $\pi_G := \frac{1}{|G|} \sum_{g \in G} g \pi g^{-1}$. Let us prove that it is a projector onto M_1, which will prove Theorem 3.1.3.

Since M_1 is G-stable, we have $\pi g \pi = g \pi$. Hence

$$\pi_G^2 = \frac{1}{|G|^2} \sum_{g,h \in G} g \pi g^{-1} h \pi h^{-1}$$

$$= \frac{1}{|G|} \sum_{g \in G} g \pi g^{-1} = \pi_G .$$

Moreover, $\pi_G(M) \subset M_1$, and for $m \in M_1$, we have

$$\pi_G(m) = \frac{1}{|G|} \sum_{g \in G} g \pi g^{-1}(m) = \frac{1}{|G|} \sum_{g \in G} g g^{-1}(m) = m .$$

hence π_G is indeed a G-stable projector onto M_1. $\square$

Remark 3.1.4 The reader can easily convince himself that what has been proved so far, from the beginning of the section, works not only for a characteristic zero field K, but also for a field k whose characteristic does not divide $|G|$.

Corollary 3.1.5 (1) *A K-representation of G is indecomposable if and only if it is irreducible.*
(2) *Every K-representation of G is a direct sum of irreducible representations.*

The second assertion of the above corollary expresses the fact that any K-representation of G is *completely reducible*.

One gets also another characterization of irreducibility.

Corollary 3.1.6 *Let S be a KG-module. The following assertions are equivalent.*

(i) *S is irreducible.*
(ii) *The K-algebra $\mathrm{End}_{KG}(S)$ is a division algebra.*

Proof (i)$\Rightarrow$(ii) is assertion (2) of Schur's Lemma 2.3.9.

(ii)$\Rightarrow$(i): if S were not irreducible, by Maschke's Theorem 3.1.3 the algebra $\mathrm{End}_{KG}(S)$ would contain an idempotent different from 0 and 1, which is impossible if it is a field. $\square$

The next corollary will be useful for character theory.

Corollary 3.1.7 *Let $x \in KG$. Assume that for all irreducible representations (S, ρ_S) of G we have $\rho_S(x) = 0$. Then $x = 0$.*

Proof Let $\rho_{KG} : KG \to \text{End}_K(KG)$ denote the morphism associated with the regular representation. Since that regular representation is a direct sum of irreducible representations, the hypothesis shows that $\rho_{KG}(x) = 0$. This implies $x = 0$ by Lemma 2.3.19.   □

3.1.3 Spectrum

Let us denote by e_G the lcm of the orders of the elements of G.

Notice that if (M, ρ) is a representation of G, then for all $g \in G$, $\rho(g)^{e_G} - 1 = 0$. Let K_G denote a splitting field of this polynomial. Then this polynomial has simple roots over K_G. The next proposition follows.

Proposition 3.1.8 *Assume that (M, ρ) is a degree r K-representation of G. For all $g \in G$,*

(1) *$\rho(g)$ is diagonalisable over K_G,*
(2) *the spectrum of $\rho(g)$ consists of $|g|$-th roots of unity (where $|g|$ denotes the order of g).*

3.2 Characters

3.2.1 First Properties

Generalities.
If (M, ρ) is a K-representation of G (resp. if M is a KG-module) its *character* is the central function $\chi_M : G \to K$ defined by

$$\chi_M(g) := \text{tr}_{M/K}\rho(g) \quad (\text{resp. } \chi_M(g) := \text{tr}_{M/K}(m \mapsto gm)).$$

The character of the trivial representation of G will be denoted by 1_G, or even 1 when there is no ambiguity. Thus

$$1_G(g) = 1 \quad \text{for all } g \in G.$$

Proposition 3.2.1 *Let M and N be KG-modules.*

(1) *$\chi_M(1) = [M : K]$.*
(2) *$\chi_{M \oplus N} = \chi_M + \chi_N$.*
(3) *$\chi_{M \otimes_K N} = \chi_M \chi_N$, i.e., $\forall g \in G$, $\chi_{M \otimes_K N}(g) = \chi_M(g)\chi_N(g)$.*

(4) $\forall g \in G$, $\chi_{M^*}(g) = \chi_M(g^{-1})$.

(5) $\chi_{\mathrm{Hom}_K(M,N)} = \chi_{M^*} \chi_N$, i.e., *for all* $g \in G$, $\chi_{\mathrm{Hom}_K(M,N)}(g) = \chi_M(g^{-1}) \chi_N(g)$.

(6) *If* $M \cong N$, *then* $\chi_M = \chi_N$.

Proof (1) and (2) are trivial.

(3) is an immediate consequence of Corollary 1.2.12.

(4) follows from Proposition 2.3.8.

(5) is an application of Proposition 3.1.8.

(6) is left to the reader. $\qquad\square$

We call *kernel of* χ_M and we denote $\ker(\chi_M)$ the kernel of the underlying morphism $\rho : G \to GL(M)$.

Corollary 3.2.2 *Let M be a KG-module with character χ_M and let $g \in G$. The following assertions are equivalent.*

(i) $g \in \ker(\chi_M)$,

(ii) $\chi_M(g) = \chi_M(1)$.

Proof Indeed, if $(\zeta_1, \ldots, \zeta_r)$ denotes the spectrum of g in its action on M, we have $|\zeta_i| = 1$ and $\chi_M(g) = \zeta_1 + \cdots + \zeta_r$, hence

$$\begin{cases} |\chi_M(g)| \leq r, \\ \chi_M(g) = r \text{ if and only if } (\forall i)\, \zeta_i = 1. \end{cases}$$

$\qquad\square$

An application: normal subgroups and characters.

We denote by $\mathrm{Irr}_K(G)$ a complete set of representatives of irreducible KG-modules up to isomorphisms.

We shall see that the normal subgroups of G may be described as the intersections of the kernels of characters of suitable irreducible KG-modules.

If H is a normal subgroup of G, whenever $\rho : G/H \to GL(V)$ is a representation (resp. $\chi : G/H \to K$ is the character of a representation) of G/H, the composition

$$G \twoheadrightarrow G/H \to GL(V) \quad (\text{resp. } G \twoheadrightarrow G/H \to K)$$

defines a representation (resp. the character of a representation) of G which is irreducible if and only if ρ is irreducible. We denote by

$$\mathrm{Infl}^G_{G/H} : \mathrm{Irr}_K(G/H) \longrightarrow \mathrm{Irr}_K(G)$$

and call *inflation* the corresponding injection (and similarly for characters).

Proposition 3.2.3 *The normal subgroups of G are the intersections of the kernels of families of irreducible characters.*

More precisely, let H be a normal subgroup of G. Then

$$H = \bigcap_{S \in \mathrm{Infl}^G_{G/H}(\mathrm{Irr}_K(G/H))} \ker \chi_S.$$

Proof It is clear that H is contained in the kernel of any character inflated from G/H. The converse inclusion follows from the fact (see Corollary 3.1.7) that the intersection of the kernels of all irreducible characters of G/H is trivial.   □

Extension and restriction of scalars.
Assume that K is a subfield of L.
    The following lemma is obvious.

Lemma 3.2.4 *Let M be a KG-module. Let χ_{LM} denote the character of the LG-module $L \otimes_K M$.*
    Then, as central functions on G with values in L,

$$\chi_{LM} = \chi_M.$$

Assume moreover that $[L : K]$ is finite. For $\lambda \in L$, we denote by $\mathrm{tr}_{L/K}(\lambda)$ the trace of the K-linear endomorphism of L defined by the multiplication by λ (see above Lemma 1.2.19).

Lemma 3.2.5 *Assume that $[L : K]$ is finite. Let N be an LG-module with character χ_N. Let N_K be the KG-module defined by considering the natural structure of K-vector space of N, and denote by χ_{N_K} its character. Then*

$$\chi_{N_K} = \mathrm{tr}_{L/K} \cdot \chi_N.$$

Proof This is an immediate consequence of Lemma 1.2.19.   □

Characters of symmetric and alternating squares.
For notation used in the following proposition, we refer the reader to Definition 1.3.1.

Proposition 3.2.6 *Let (M, ρ) be a representation of G. Let us denote by*

- χ^s_M *the character of* $\mathrm{Sym}^2(M)$,
- χ^a_M *the character of* $\mathrm{Alt}^2(M)$.

Then, for all $g \in G$,

$$\begin{cases} \chi^s_M(g) = \frac{1}{2}\big(\chi_M(g)^2 + \chi_M(g^2)\big) \\ \chi^a_M(g) = \frac{1}{2}\big(\chi_M(g)^2 - \chi_M(g^2)\big). \end{cases}$$

Proof Up to extending the scalars to K_G (the splitting field of $X^{e_G} - 1$, see above Proposition 3.1.8), which we do, we may assume that for all $g \in G$, $\rho(g)$ is diagonalizable.

For $g \in G$, choose a basis $(e_i)_{1 \leq i \leq r}$ of M consisting of eigenvectors of $\rho(g)$. For all i $(1 \leq i \leq r)$, let us set $\rho(g)(e_i) = \lambda_i e_i$.

Using the basis $(e_i \otimes_K e_j + e_j \otimes_K e_i)_{i \leq j}$ for $\mathrm{Sym}^2(M)$ (see Proposition 1.3.2, we see that

$$\chi_M^s(g) = \sum_{i<j} \lambda_i \lambda_j + \sum_i \lambda_i^2 = \frac{1}{2}\left(\left(\sum_i \lambda_i\right)^2 + \sum_i \lambda_i^2\right)$$
$$= \frac{1}{2}\left(\chi_M(g)^2 + \chi_M(g^2)\right).$$

Now since $M \otimes_K M = \mathrm{Sym}^2(M) \oplus \mathrm{Alt}^2(M)$, we see that

$$\chi_M^a(g) = \chi_{M \otimes_K M}(g) - \chi_M^s(g) = \chi_M(g)^2 - \chi_M^s(g)$$
$$= \frac{1}{2}\left(\chi_M(g)^2 - \chi_M(g^2)\right).$$

$\square$

Exercise 3.2.7 Let M and N be KG-modules. Prove that

$$(\chi_M + \chi_N)^s = \chi_M^s + \chi_N^s + \chi_M \chi_N,$$
$$(\chi_M + \chi_N)^a = \chi_M^a + \chi_N^a + \chi_M \chi_N.$$

3.2.2 Orthogonality Relations and First Applications

The space $\mathrm{CF}(G, \mathbb{Q}_K^G)$.
We recall that K is a characteristic zero field and G is a finite group, and that e_G denotes the lcm of the orders of the elements of G.

Notation 3.2.8 (*Notation for fields*)

(1) Let $\mu_{e_G}(K)$ denote the group of e_G-roots of unity of a splitting field over K of $X^{e_G} - 1$ (a field we have denoted by K_G).
(2) Since $\mathbb{Q}$ is the prime field of K, the field $\mathbb{Q}(\mu_{e_G}(K))$ is a subfield of $K(\mu_{e_G}(K))$. We denote by $*$ the automorphism of $\mathbb{Q}(\mu_{e_G}(K))$ which sends an e_G-th root of unity ζ to its inverse ζ^{-1}.

Notation 3.2.9 (*Notation for fields of characters*)

(1) If M is a KG-module, we denote by $\mathbb{Q}(\chi_M)$ the subfield of K generated by the family $(\chi_M(g))_{g \in G}$.
(2) We denote by $\mathbb{Q}_K^G$ the subfield of K generated by all fields $\mathbb{Q}(\chi_M)$ for all KG-modules M.

Remarks 3.2.10 (1) For any embedding $K \subset \mathbb{C}$, $\mathbb{Q}(\boldsymbol{\mu}_{e_G}(K)) \subset \mathbb{C}$, and the automorphism $*$ is nothing but the restriction to $\mathbb{Q}(\boldsymbol{\mu}_{e_G}(K))$ of the complex conjugation.

(2) $\mathbb{Q}_K^G$ is a subfield of both K and $\mathbb{Q}(\boldsymbol{\mu}_{e_G}(K))$.

(3) All fields $\mathbb{Q}(\chi_M)$, hence the field $\mathbb{Q}_K^G$, are stable by the automorphism $*$.

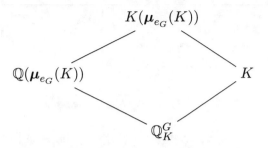

Notation 3.2.11 (*Notation for class functions*)

(1) We recall (see Convention 2.3.5) that $CF(G, \mathbb{Q}_K^G)$ denotes the $\mathbb{Q}_K^G$-vector space of class functions from G to $\mathbb{Q}_K^G$.

(2) We define a hermitian product (relative to $*$) on $CF(G, \mathbb{Q}_K^G)$ by the formula

$$\langle \alpha_1, \alpha_2 \rangle_G := \frac{1}{|G|} \sum_{g \in G} \alpha_1(g)\alpha_2(g)^* \quad \text{for } \alpha_1, \alpha_2 \in CF(G, \mathbb{Q}_K^G).$$

When the context is clear, we may omit the subscript G and write $\langle \alpha_1, \alpha_2 \rangle := \langle \alpha_1, \alpha_2 \rangle_G$.

Note that the above hermitian product is not defined in general on $CF(G, K)$, since we make no hypothesis whatsoever about K (except that it has characteristic zero).

The following lemma is straightforward.

Lemma 3.2.12 (1) *The $\mathbb{Q}_K^G$-vector space $CF(G, \mathbb{Q}_K^G)$ has as basis the family of characteristic functions $(\gamma_C)_{C \in Cl(G)}$ of conjugacy classes of G.*

(2) *$(\gamma_C)_{C \in Cl(G)}$ is an orthogonal basis of $CF(G, \mathbb{Q}_K^G)$. More precisely, for $C, D \in Cl(G)$, we have*

$$\langle \gamma_C, \gamma_D \rangle_G = \delta_{C,D} \frac{|C|}{|G|}.$$

Let M be a KG-module. Then its character χ_M belongs to the space $CF(G, \mathbb{Q}_K^G)$. We have

$$\chi_M = \sum_{C \in Cl(G)} \chi_M(g_C)\gamma_C$$

(where g_C denotes an element of C).

Remark 3.2.13 For χ_M the character of a KG-module M, we have (see Proposition 3.2.1, (5)) $\chi_{M^*}(g) = \chi_M(g^{-1})$. It follows that

$$\chi_{M^*}(g) = \chi_M(g)^*.$$

Hence for KG-modules M and N,

$$\langle \chi_M, \chi_N \rangle_G = \frac{1}{|G|} \sum_{g \in G} \chi_M(g)\chi_N(g)^* = \frac{1}{|G|} \sum_{g \in G} \chi_M(g)\chi_N(g^{-1})$$

$$= \frac{1}{|G|} \sum_{g \in G} \chi_M(g^{-1})\chi_N(g) = \frac{1}{|G|} \sum_{g \in G} \chi_M(g)^*\chi_N(g)$$

$$= \langle \chi_N, \chi_M \rangle_G.$$

Notation 3.2.14 (*The commuting field*) If S is an irreducible KG-module, we set

$$D_S := \mathrm{End}_{KG}(S).$$

Recall that by Schur's lemma (Proposition 2.3.9) D_S is a division K-algebra, a finite extension of K.

Proposition 3.2.15 *Let* $\mathrm{Irr}_K(G)$ *be a complete set of representatives of isomorphism classes of irreducible K-representations of G.*

(1) *For* $S, T \in \mathrm{Irr}_K(G)$,

$$\big[\mathrm{Hom}_{KG}(S, T) : K\big] = \langle \chi_S, \chi_T \rangle_G,$$

hence

$$\langle \chi_S, \chi_T \rangle_G = \begin{cases} [D_S : K] & \text{if } T - S, \\ 0 & \text{if } T \neq S. \end{cases}$$

(2) *The family* $(\chi_S)_{S \in \mathrm{Irr}_K(G)}$ *is an orthogonal system in the vector space* $\mathrm{CF}(G, \mathbb{Q}_K^G)$. *In particular it is a linearly independent system.*

Proof • For $S, T \in \mathrm{Irr}_K(G)$, we know by Lemma 2.1.4 that

$$\mathrm{Hom}_{KG}(S, T) = \mathrm{Fix}^G\big(\mathrm{Hom}_K(S, T)\big).$$

• By Proposition 3.2.1, (4), we know that the character of the KG-module $\mathrm{Hom}_K(S, T)$ is $\chi_S^* \chi_T$.
• Then it follows from Lemma 3.1.1, (3), that

$$\big[\mathrm{Hom}_{KG}(S, T) : K\big] = \frac{1}{|G|} \sum_{g \in G} \chi_S(g)^* \chi_T(g) = \langle \chi_S, \chi_T \rangle_G,$$

which proves (1).

(2) For $S, T \in \mathrm{Irr}_K(G)$, $S \neq T$, it follows from Schur's Lemma Proposition 2.3.9, (1), that $\mathrm{Hom}_{KG}(S, T) = 0$, hence (2) (by the preceding assertion).   □

Note that, as a particular case of Proposition 3.2.15, (1), we have

$$\left[\mathrm{End}_{KG}(S) : K\right] = \langle \chi_S, \chi_S \rangle_G .$$

Characters characterize representations.

Proposition 3.2.15 has important applications to the classification of representations, as we shall see now.

First of all, Proposition 3.2.15, (2), states that the family $(\chi_S)_{S \in \mathrm{Irr}_K(G)}$ is a free system in $\mathrm{CF}(G, \mathbb{Q}_K^G)$. Since that space is finite dimensional (its dimension is $|\mathrm{Cl}(G)|$ by Lemma 2.3.17), the next proposition follows.

Proposition 3.2.16 *The set* $\mathrm{Irr}_K(G)$ *is finite. More precisely*

$$|\mathrm{Irr}_K(G)| \leq |\mathrm{Cl}(G)| .$$

Remark 3.2.17 We shall prove later (Theorem 3.5.27) a much more precise result.

Proposition 3.2.18 *Let M be a KG-module. Denote by*

$$M = \bigoplus_{i \in I} S_i \tag{Dec}$$

a decomposition of M into a direct sum of irreducible KG-modules.

(1) *For all $S \in \mathrm{Irr}_K(G)$, the number of i such that $S_i \cong S$ is equal to $\langle \chi_M, \chi_S \rangle_G / \langle \chi_S, \chi_S \rangle_G$.*

(2) *In particular this number does not depend on the choice of the decomposition (Dec).*

Proof Indeed, let us denote by $m_{S,\mathrm{Dec}}$ the number of i such that $S_i \cong S$. Since $\chi_M = \sum_i \chi_{S_i}$, we have $\chi_M = \sum_S m_{S,\mathrm{Dec}} \chi_S$, which implies Proposition 3.2.18.   □

Notation 3.2.19 The number of i such that $S_i \cong S$ in a decomposition as (Dec) above is called the *multiplicity of S in M* and denoted by $m_{S,M}$.

Note that the above corollary proves that $\langle \chi_S, \chi_S \rangle_G$ divides $\langle \chi_M, \chi_S \rangle_G$ – see Exercise 3.7.18 for a direct proof (not using characters).

ⓘ **Attention** ⓘ By no way is the decomposition of a KG-module into a direct sum of irreducible submodules unique. For example, one may consider the case where $G = 1$. Then, for a KG-module M, such decompositions correspond to choosing a decomposition of M into a direct sum of one-dimensional subspaces.

Remark 3.2.20 Any KG-module M such that $\langle \chi_M, \chi_M \rangle_G = 1$ is irreducible.

(!) The converse is wrong in general. Consider for example $K = \mathbb{Q}$, $G = \mu_3 = \{1, \zeta, \zeta^2\}$ where $\zeta = \exp(2i\pi/3)$, and $M = \mathbb{Q}(\zeta)$, a 2-dimensional K-vector space. Then M is an irreducible KG-module, but $\langle \chi_M, \chi_M \rangle_G = 2$.

Theorem 3.2.21 *Let M and N be KG-modules. The following assertions are equivalent.*

(i) $M \cong N$.

(ii) $\chi_M = \chi_N$.

(iii) *For all $S \in \mathrm{Irr}_K(G)$, $m_{S,M} = m_{S,N}$.*

Remark 3.2.22 Notice that the equivalence between (i) and (ii) in the above theorem may be phrased as follows (the reader is invited to think about the strength of that statement).

Let K be a characteristic zero field. Let G be a finite group, and let $\rho_i : G \to \mathrm{GL}_r(K)$ be group morphisms for $i = 1, 2$. Then the following assertions are equivalent

(i) There exists $U \in \mathrm{GL}_r(K)$ such that

$$\forall g \in G, \ \rho_2(g) = U\rho_1(g)U^{-1}.$$

(ii) For all $g \in G$, $\mathrm{tr}\rho_1(g) = \mathrm{tr}\rho_2(g)$.

Proof of Theorem 3.2.21.

(i) $\Rightarrow$(ii) is Proposition 3.2.1, (6).

(ii) $\Rightarrow$(iii) results from Proposition 3.2.18.

(iii) $\Rightarrow$(i) follows from the isomorphism

$$M \cong \bigoplus_{S \in \mathrm{Irr}_K(G)} S^{m_{S,M}}.$$

□

A first (immediate, but in a sense spectacular) consequence of the preceding characterization of a representation (up to isomorphism) by its character is the following corollary.

Corollary 3.2.23 *A linear representation of a finite group over a characteristic zero field is characterized (up to isomorphism) by its restrictions to cyclic subgroups.*

Corollary 3.2.24 *Let M and N be KG-modules. The following assertions are equivalent.*

(i) $M \cong N$.

(ii) *There exists a field extension L of K and a natural integer $r \geq 1$ such that*
$(LM)^r \cong (LN)^r$.

Proof (i)$\Rightarrow$(ii) is trivial. Let us prove (ii)$\Rightarrow$(i).

If $(LM)^r \cong (LN)^r$, then $\chi_{(LM)^r} = \chi_{(LN)^r}$, i.e., $r\chi_{LM} = r\chi_{LN}$ and $\chi_{LM} = \chi_{LN}$. Since $\chi_{LM} = \chi_M$, we see that $\chi_M = \chi_N$ and $M \cong N$. $\square$

Regular representation.

Let us apply what precedes to the case of the regular representation.

We denote by χ_G^{reg} the character of the regular representation of G, i.e., of the KG-module KG where G acts by left multiplication.

The proof of the next lemma is immediate (see also the proof of Lemma 3.2.38 below in a more general context).

Lemma 3.2.25 $\chi_G^{\mathrm{reg}}(g) = \begin{cases} |G| & \text{if } g = 1, \\ 0 & \text{if } g \neq 1. \end{cases}$

We recall (see Notation 3.2.14) that for S an irreducible KG-module, we denote by D_S the division K-algebra (see Proposition 2.3.9, (2)) $\mathrm{End}_{KG}(S)$.

Since D_S is a subalgebra of $\mathrm{End}_K(S)$, S is naturally endowed with a structure of D_S-module, hence of D_S-vector space since D_S is a field. We denote by $[S : D_S]$ the dimension of that vector space (which is finite since S is finite dimensional over K and $K \subseteq D_S$).

Proposition 3.2.26 (1) *For $S \in \mathrm{Irr}_K(G)$,*

$$m_{S,KG} = [S : D_S],$$

hence

$$\chi_G^{\mathrm{reg}} = \sum_{S \in \mathrm{Irr}_K(G)} [S : D_S]\chi_S.$$

(2) *In particular*

  (a) $|G| = \sum_{S \in \mathrm{Irr}_K(G)} [\mathrm{End}_{D_S}(S) : K]$.
  (b) *Every irreducible representation of KG occurs in the regular representation.*

Proof (1) We have

$$m_{S,KG} = \frac{\langle \chi_G^{\mathrm{reg}}, \chi_S \rangle}{[D_S : K]} = \frac{[S : K]}{[D_S : K]} = [S : D_S].$$

(2) By (1), we have

$$|G| = \sum_{S \in \mathrm{Irr}_K(G)} [S : D_S][S : K].$$

Now

$$[\mathrm{End}_{D_S}(S) : K] = [\mathrm{End}_{D_S}(S) : D_S][D_S : K]$$
$$= [S : D_S]^2[D_S : K] = [S : D_S][S : K].$$

$\square$

An example: Galois extension.

Let L be a field, an extension of K such that L/K is a finite Galois extension with Galois group G. Then L may be viewed as a KG-module.

Item (2) of the following proposition is called the "normal basis theorem".

Theorem 3.2.27 *Let L/K be a finite Galois extension with Galois group G.*

(1) *As a KG-module, L is isomorphic to the regular module KG.*
(2) *There exists an element $x \in L$ such that $(gx)_{g \in G}$ is a K-basis of L.*

Proof (2) is an immediate consequence of (1). We prove (1).

Let us denote by L^* the K-dual of L, endowed with the contragredient action of the Galois group G.

• Let us denote by M the KG-module defined from L^* by extending scalars from K to L, and then restricting the scalars from L to K. Thus M is the KG-module $L \otimes_K L^*$, where the action of KG is given by

$$\lambda g.(x \otimes y) := x \otimes \lambda y g^{-1} \quad \text{for } \lambda \in K \text{ and } g \in G,$$

isomorphic, as a KG-module, to $(L^*)^{[L:K]}$.

• View $\mathrm{End}_K(L)$ as endowed with the structure of KG-module defined by $g \cdot f := fg^{-1}$ for $g \in G$ and $f \in \mathrm{End}_K(L)$. Then the map

$$M \to \mathrm{End}_K(L) , \quad x \otimes_K y \mapsto (\mu \mapsto y(\mu)x),$$

defines an isomorphism of KG-modules.

• Left multiplication by the elements of L endows $\mathrm{End}_K(L)$ with a structure of L-vector space. Since $[\mathrm{End}_K(L) : K] = [L : K]^2$, we have $[\mathrm{End}_K(L) : L] = [L : K]$.

It is known (this is "Dedekind Lemma", see for example [Bro13, Proposition 2.249] that the family of elements of G (viewed as elements of $\mathrm{End}_K(L)$) is free over L. Since $|G| = [L : K]$, it follows that G is a basis of $\mathrm{End}_K(L)$ as an L-vector space.

The action of KG on $\mathrm{End}_K(L)$ described above can then be also described as follows. For $s \in G$ and $\sum_{g \in G} \lambda_g g \in \mathrm{End}_K(L)$ ($\lambda_g \in L$), we have

$$s \cdot \sum_{g \in G} \lambda_g g = \sum_{g \in G} \lambda_g g s^{-1} \in \mathrm{End}_K(L).$$

Notice that, chosen an element $e \in L$ ($e \neq 0$), the KG-submodule $KeG = \bigoplus_{g \in G} Keg$ of $\mathrm{End}_K(L)$ is isomorphic to KG where the elements $s \in G$ act by right multiplication by s^{-1}, hence is isomorphic to the contragredient KG-module $(KG)^*$. Thus, choosing a basis (e_i) of L over K, we have the following isomorphism of KG-modules

$$\mathrm{End}_K(L) = \bigoplus_i Ke_iG \simeq (KG^*)^{[L:K]} .$$

This shows that, as KG-modules,

$$(L^*)^{[L:K]} \simeq (KG^*)^{[L:K]} ,$$

and by Corollary 3.2.24, this implies

$$L \simeq KG \quad \text{as } KG\text{-modules} .$$

$\square$

3.2.3   Splitting Fields

Absolutely irreducible module or character.

Proposition-Definition 3.2.28 *Let $S \in \mathrm{Irr}_K(G)$. The following assertions are equivalent.*

(i) $D_S = K\mathrm{Id}_S$.

(ii) $\langle \chi_S, \chi_S \rangle = 1$.

(iii) *For each extension L of K, $L \otimes_K S$ is an irreducible LG-module.*

(iv) *For each finite extension L of K, $L \otimes_K S$ is an irreducible LG-module.*

If the above assertions are satisfied, we say that S is absolutely irreducible *and that χ_S is an* absolutely irreducible character.

Proof (i)$\Rightarrow$(ii) results from the formula $\langle \chi_S, \chi_S \rangle = [D_S : K]$.

(ii)$\Rightarrow$(iii) is immediate.

(iii)$\Rightarrow$(iv) is obvious.

(iv)$\Rightarrow$(i) Let $\alpha \in D_S = \mathrm{End}_{KG}(S)$. There exists a finite extension L of K in which the characteristic polynomial of α has at least one root, say λ. Since $L \otimes_K S$ is an irreducible LG-module and since $\ker(\alpha - \lambda\mathrm{Id})$ is nonzero, we have $\alpha = \lambda\mathrm{Id}$. But then $\lambda \in K$, and thus $D_S = K\mathrm{Id}_S$. $\square$

Invariant bilinear forms on absolutely irreducible modules.

Definition 3.2.29 Whenever χ is an absolutely irreducible character of G, we denote by $\nu_2(\chi)$ and we call *Frobenius–Schur indicator* the number

$$\nu_2(\chi) := \frac{1}{|G|} \sum_{g \in G} \chi(g^2).$$

We shall see now, in particular, that $\nu_2(\chi) \in \{-1, 0, 1\}$.

Proposition 3.2.30 *Let S be an absolutely irreducible KG-module with character χ_S.*

- *For each of the following assertions (n), (s) or (a), the corresponding statements (i) and (ii) are equivalent.*
- *One of these following assertions (n), (s) or (a) is true.*

(n) (i) $\nu_2(\chi_S) = 0$.
　　(ii) *There is no G-invariant bilinear form on S.*
(s) (i) $\nu_2(\chi_S) = 1$.
　　(ii) *There is a line of G-invariant bilinear forms on S, all of which are symmetric, and all but 0 are nondegenerate.*
(a) (i) $\nu_2(\chi_S) = -1$.
　　(ii) *There is a line of G-invariant bilinear forms on S, all of which are alternating, and all but 0 are nondegenerate.*

Proof A G-invariant bilinear form on S may be viewed as an element of Fix^G $(\mathrm{Hom}_K(S, S^*))$, and $\mathrm{Hom}_K(S, S^*) \simeq S^* \otimes_K S^*$. By Definition 1.3.1,

$$\mathrm{Fix}^G(S^* \otimes_K S^*) = \mathrm{Fix}^G(\mathrm{Sym}^2(S^*)) \oplus \mathrm{Fix}^G(\mathrm{Alt}^2(S^*)),$$

and $\mathrm{Fix}^G(\mathrm{Sym}^2(S^*))$ (resp. $\mathrm{Fix}^G(\mathrm{Alt}^2(S^*))$) is identified with the space of G-invariant symmetric bilinear forms on S (resp. the space of G-invariant alternating bilinear forms on S).

On the one hand, the equality

$$[\mathrm{Fix}^G(S^* \otimes_K S^*) : K] = [\mathrm{Fix}^G(\mathrm{Sym}^2(S^*)) : K] + [\mathrm{Fix}^G(\mathrm{Alt}^2(S^*)) : K]$$

translates into (see Proposition 3.2.6 for the notation)

$$\langle (\chi_S^*)^2, 1 \rangle = \langle \chi_S^*, \chi_S \rangle = \langle \chi_{S^*}^s, 1 \rangle + \langle \chi_{S^*}^a, 1 \rangle,$$

where 1 denotes the character of the trivial representation of G. Note that $\langle \chi_{S^*}^s, 1 \rangle$ and $\langle \chi_{S^*}^a, 1 \rangle$ belong to $\mathbb{N}$.

On the other hand, it follows from Proposition 3.2.6 that

$$\langle \chi_{S^*}^s, 1 \rangle = \frac{1}{2} \left(\langle (\chi_S^*)^2, 1 \rangle + \nu_2(\chi_S) \right) ,$$

$$\langle \chi_{S^*}^a, 1 \rangle = \frac{1}{2} \left(\langle (\chi_S^*)^2, 1 \rangle - \nu_2(\chi_S) \right) .$$

Hence

(n) either $S \not\simeq S^*$, that is, $\chi_S \neq \chi_S^*$, hence $\langle (\chi_S^*)^2, 1 \rangle = 0$ and $\langle \chi_{S^*}^s, 1 \rangle = \langle \chi_{S^*}^a, 1 \rangle = 0$, which implies $\nu_2(\chi_S) = 0$,

(sa) or $S \simeq S^*$, that is, $\chi_S = \chi_S^*$, hence $\langle (\chi_S^*)^2, 1 \rangle = 1$ and

    (s) either $\langle \chi_{S^*}^s, 1 \rangle = 1$ and $\nu_2(\chi_S) = 1$,

    (a) or $\langle \chi_{S^*}^a, 1 \rangle = 1$ and $\nu_2(\chi_S) = -1$.

                                                                       □

Splitting fields.

Definition 3.2.31 Let G be a finite group and let K be a characteristic zero field.

(1) Given $S \in \text{Irr}_K(G)$, we say that K is a *splitting field for* S if S is absolutely irreducible.

(2) We say that K is a *splitting field for* G if, whenever $S \subset \text{Irr}_K(G)$, S is absolutely irreducible.

Given a field K, the algebraic closure of K is an extension of K which is a splitting field for G (for example by Proposition-Definition 3.2.28). But in general that algebraic closure is much too big, as shown by the next lemma.

Lemma 3.2.32 *Given a finite group G and a field K, there is a finite extension of K which is a splitting field for G.*

Proof Let $\bar{K}$ be an algebraic closure of K. For each $S \in \text{Irr}_{\bar{K}}(G)$, choose a $\bar{K}$-basis of S, hence, for all $g \in G$, a matrix $A_S(g)$ for the action of g on S. Let us denote by L the field generated over K by all the entries of all matrices $A_S(g)$ for all $S \in \text{Irr}_{\bar{K}}(G)$ and $g \in G$.

Since L is generated over K by a finite number of algebraic elements, it has finite dimension as a K-vector space. By construction, for all $S \in \text{Irr}_{\bar{K}}(G)$, S is rational over L, hence the extension of scalars from L to $\bar{K}$ induces a natural bijection $\text{Irr}_L(G) \xrightarrow{\sim} \text{Irr}_{\bar{K}}(G)$. Whenever $S \in \text{Irr}_L(G)$ we have $\langle \chi_S, \chi_S \rangle = 1$. This shows that L is a splitting field for G.   □

Remark 3.2.33 We shall show later a much more precise result, due to Richard Brauer (see Theorem 6.2.4): if e_G denotes the lcm of the orders of elements of G, then the field $\mathbb{Q}(\mu_{e_G})$ is a splitting field for G (hence for all its subgroups as well).

The following fundamental property of splitting fields is an immediate consequence of the definition.

Proposition 3.2.34 *Let K be a splitting field for G, and let L be an extension of K.*

(1) *L is a splitting field for G.*
(2) *The extension of scalars from K to L induces a bijection*

$$\mathrm{Irr}_K(G) \overset{\sim}{\longrightarrow} \mathrm{Irr}_L(G).$$

Previous results restated over a splitting field.

Let us first revisit Schur's Lemma 2.3.9 when K is a splitting field.

Lemma 3.2.35 *Let S be an irreducible G-representation. Assume K is a splitting field for G. Then for all $\alpha \in \mathrm{End}_K(S)$,*

$$\frac{1}{|G|} \sum_{g \in G} g \alpha g^{-1} = \frac{\mathrm{tr}(\alpha)}{[S:K]} \mathrm{Id}_S.$$

Proof By Schur's Lemma (Corollary 3.1.2, (2)), $\dfrac{1}{|G|} \sum_{g \in G} g \alpha g^{-1} = \lambda \mathrm{Id}_S$ for some $\lambda \in K$. Computing traces of the two sides of the preceding equality gives the answer. $\square$

Let us now examine the orthogonality relations. If K is a splitting field for G, then for all $S \in \mathrm{Irr}_K(G)$ we have $\langle \chi_S, \chi_S \rangle = 1$ (for example by item (ii) of Proposition-Definition 3.2.28). This makes the next proposition obvious.

Proposition 3.2.36 *Assume K is a splitting field for G.*

(1) *The family $(\chi_S)_{S \in \mathrm{Irr}_K(G)}$ is an orthonormal family in the space $\mathrm{CF}(G, K)$.*
(2) *For M a KG-module,*

$$\langle \chi_M, \chi_M \rangle_G = \sum_{S \in \mathrm{Irr}_K(G)} m_{S,M}^2.$$

(3) *One has $\langle \chi_M, \chi_M \rangle_G = 1$ if and only if M is irreducible.*

Now we examine the case of the regular representation. The next proposition is an immediate consequence of Proposition 3.2.26.

Proposition 3.2.37 *Assume that K is a splitting field for G.*

(1) *For $S \in \mathrm{Irr}_K(G)$,*

$$m_{S,KG} = \chi_S(1),$$

hence

$$\chi_G^{\mathrm{reg}} = \sum_{S \in \mathrm{Irr}_K(G)} \chi_S(1) \chi_S = \sum_{S \in \mathrm{Irr}_K(G)} [S:K] \chi_S.$$

(2) *In particular*

$$|G| = \sum_{S \in \mathrm{Irr}_K(G)} \chi_S(1)^2 = \sum_{S \in \mathrm{Irr}_K(G)} [S : K]^2 .$$

3.2.4 From Permutations to Characters

Let X be a finite G-set. Then the $\mathbb{Q}$-vector space $\mathbb{Q}X$ with basis X inherits naturally a structure of $\mathbb{Q}G$-module (see Sect. 2.3.1), with character denoted by χ_X. We still denote by 1 the trivial character (character of the trivial representation) of G.

Lemma 3.2.38 *With the above notation:*

(1) *for all $g \in G$, $\chi_X(g) = |\mathrm{Fix}^{\langle g \rangle}(X)|$,*
(2) *$\langle \chi_X, 1 \rangle$ is equal to the number of orbits of G on X,*
(3) *$\chi_{X \times X} = \chi_X^2$.*

Proof (1) The matrix representing the linear action of g on the basis X is monomial, and nonzero diagonal entries correspond to fixed points under g. This make the first assertion clear.
(2) By Lemma 3.1.1, we know that

$$\langle \chi_X, 1 \rangle = [\mathrm{Fix}^G(\mathbb{Q}X) : \mathbb{Q}] \quad \text{and} \quad \mathrm{Fix}^G(\mathbb{Q}X) = \pi_{\mathbb{Q}X}^G(\mathbb{Q}X) .$$

Moreover, if $G \backslash X$ denotes the set of orbits of X under G, it is clear that

$$\mathbb{Q}X = \bigoplus_{\Omega \in G \backslash X} \mathbb{Q}\Omega ,$$

and for all Ω,

$$\pi_{\mathbb{Q}X}^G(\mathbb{Q}\Omega) = \mathbb{Q}S\Omega \quad \text{where } S\Omega = \sum_{w \in \Omega} w .$$

It follows that

$$\mathrm{Fix}^G(\mathbb{Q}X) = \bigoplus_{\Omega \in G \backslash X} \mathbb{Q}S\Omega ,$$

proving in particular assertion (2).
(3) results from the fact that $\chi_{X \times X} = \chi_{\mathbb{Q}X \otimes_{\mathbb{Q}} \mathbb{Q}X}$.   □

Notice that $X \times X$ is *never* a transitive G-set (unless $|X| \le 1$). Indeed, if

$$\Delta(X) := \{(x, x) \mid x \in X\}$$

denotes the diagonal of $X \times X$, and if

$$\Delta'(X) := \{(x, x') \mid (x \in X)(x' \in X)(x \ne x')\}$$

denotes its complement in $X \times X$, we have the G-stable partition

$$X \times X = \Delta(X) \sqcup \Delta'(X).$$

Proposition-Definition 3.2.39 *Let X be a G-set. We assume that $|X| \ge 2$. The following assertions are equivalent.*

(i)   *G has exactly two orbits on $X \times X$.*
(ii) *G is transitive on X and on $\Delta'(X)$.*
(iii) *G is transitive on X and for $x \in X$, its fixator G_x is transitive on $X \setminus \{x\}$.*

If the above assertions are true, we say that G is doubly transitive *on X.*

Proof The equivalence of (i) and (ii) is immediate, as well as the implication (ii)$\Rightarrow$(iii).

Now assume (iii). Let $(x, x'), (y, y') \in \Delta'(X)$. Since G is transitive on X, there is $g \in G$ such that $y = gx$. Note that $gx' \ne y$. Since G_y is transitive on $X \setminus \{y\}$ and $y' \ne y$, there exists $h \in G_y$ such that $y' = hgx'$. Then we see that $hg(x, x') = (y, y')$, and this shows (ii). $\qquad\qquad\square$

Proposition 3.2.40 *Assume that $|X| \ge 2$, and that G is transitive on X. The following assertions are equivalent.*

(i)   *G is doubly transitive on X.*
(ii) *$\langle \chi_X^2, 1 \rangle = 2$.*
(iii) *$\langle \chi_X - 1, \chi_X - 1 \rangle = 1$, i.e., $\chi_X - 1$ is the character of an absolutely irreducible $\mathbb{Q}G$-module.*

Proof (i)$\Leftrightarrow$(ii) follows from Proposition 3.2.39.
(ii)$\Leftrightarrow$(iii): follows from Lemma 3.2.38 above.
Now note that

$$\langle \chi_X - 1, \chi_X - 1 \rangle = \langle \chi_X, \chi_X \rangle - 2\langle \chi_X, 1 \rangle + 1,$$

and since χ_X is $\mathbb{Z}$-valued,

$$\langle \chi_X - 1, \chi_X - 1 \rangle = \langle \chi_X^2, 1 \rangle - 2\langle \chi_X, 1 \rangle + 1.$$

Now since G is transitive on X, it follows from Lemma 3.2.38 above that $\langle \chi_X, 1 \rangle = 1$, hence

$$\langle \chi_X - 1, \chi_X - 1 \rangle = \langle \chi_X^2, 1 \rangle - 1 .$$

$\square$

3.3 Structure of the Group Algebra

3.3.1 A Product of Endomorphism Algebras

Note that, for an irreducible KG-module S, by definition of D_S, the image of the "structural algebra morphism" $\rho_S : KG \to \mathrm{End}_K(S)$ is contained in $\mathrm{End}_{D_S}(S)$. We still denote by

$$\rho_S : KG \to \mathrm{End}_{D_S}(S)$$

the induced algebra morphism.

Theorem 3.3.1 *The algebra morphism*

$$\prod_{S \in \mathrm{Irr}_K(G)} \rho_S : KG \to \prod_{S \in \mathrm{Irr}_K(G)} \mathrm{End}_{D_S}(S)$$

is an isomorphism.

Remark 3.3.2 The surjectivity of the morphism $\rho_S : KG \to \mathrm{End}_{D_S}(S)$ is known as the *double centralizer property*.

Indeed, it expresses the fact that "what commutes to what commutes to the image of KG in $\mathrm{End}_{D_S}(S)$ is the image of KG".

Proof The above map is clearly a K-algebra morphism. Both sides have the same dimension by Proposition 3.2.26, (2). Thus it suffices to prove that the above morphism is injective: this follows from Corollary 3.1.7. $\square$

For all S, let us denote by e_S the inverse image of the unit element of $\mathrm{End}_{D_S}(S)$ under the above isomorphism. The following property is immediate.

Lemma 3.3.3 *The family $(e_S)_{S \in \mathrm{Irr}_K(G)}$ is a family of central orthogonal idempotents, that is*

(1) $\forall S \in \mathrm{Irr}_K(G)$, $e_S \in ZKG$,
(2) $\forall S, T \in \mathrm{Irr}_K(G)$, $e_S e_T = \delta_{S,T} e_S$,
(3) $\sum_{S \in \mathrm{Irr}_K(G)} e_S = 1$.

Notice that for $S \in \mathrm{Irr}_K(G)$, KGe_S is a two-sided ideal of KG. Moreover, KGe_S is an algebra ($(!)$ if $|G| > 1$, it is not a subalgebra of KG, since its unit element is e_S), and the natural injections $KGe_S \hookrightarrow KG$ define an algebra isomorphism

$$\prod_{S \in \mathrm{Irr}_K(G)} KGe_S \xrightarrow{\sim} KG.$$

Theorem 3.3.1 can then be re-stated as follows.

Theorem 3.3.4 (1) *Whenever* $S, T \in \mathrm{Irr}_K(G)$,

$$\rho_S(e_T) = \begin{cases} \mathrm{Id}_S & \text{if } S = T, \\ 0 & \text{if } S \neq T. \end{cases}$$

(2) *For each S, the algebra morphism*

$$\rho_S : KGe_S \to \mathrm{End}_{D_S}(S)$$

is an isomorphism.

We shall now compute the explicit inverse of the above isomorphism.

Theorem 3.3.5 (Fourier inversion) *The map*

$$\begin{cases} \displaystyle\prod_{S \in \mathrm{Irr}_K(G)} \mathrm{End}_{D_S}(S) \to KG, \\[2ex] (\alpha_S)_{S \in \mathrm{Irr}_K(G)} \mapsto \dfrac{1}{|G|} \displaystyle\sum_{g \in G} \Big(\sum_{S \in \mathrm{Irr}_K(G)} [S : D_S] \mathrm{tr}_{S/K}(g^{-1}\alpha_S) \Big) g \end{cases}$$

is the inverse of the above isomorphism.

Proof We shall first prove the above formula in the case where $\alpha_S = \mathrm{Id}_S$ for all $S \in \mathrm{Irr}_K(G)$, by proving two lemmas.

Let e be an idempotent of ZKG. The KG-module KGe, where G acts by left multiplication, is a summand of the regular module KG. We denote by χ_{KGe} its character.

Lemma 3.3.6 *Let e be an idempotent of ZKG. Then*

$$\chi_{KGe}^0 = |G|e.$$

Proof of Lemma 3.3.6. For all $x \in KG$, we have $\chi_{KGe}(x) = \chi_G^{\mathrm{reg}}(xe)$. Thus it follows from Lemma 2.3.16 that $\chi_{KGe}^0 = e(\chi_G^{\mathrm{reg}})^0$. But $(\chi_G^{\mathrm{reg}})^0 = |G|1$.

Notice that the following lemma is the particular case of Theorem 3.3.5 where $\alpha_S = \mathrm{Id}_S$.

Lemma 3.3.7 *For each $S \in \mathrm{Irr}_K(G)$, the corresponding central idempotent is*

$$e_S = \frac{[S : D_S]}{|G|} \chi_S^0 = \frac{[S : D_S]}{|G|} \sum_{g \in G} \chi_S(g^{-1})g \, .$$

Proof of Lemma 3.3.7. By item (1) of Theorem 3.3.4, and item (1) of Proposition 3.2.26, we see that

$$\chi_{KGe_S} = [S : D_S]\chi_S \, .$$

Now it suffices to apply Lemma 3.3.6.

Now we can prove Theorem 3.3.5. Since $\rho_S : KGe_S \to \mathrm{End}_{D_S}(S)$ is an isomorphism, and since the formula given in Theorem 3.3.5 is linear in α_S, we may assume that $\alpha_S = \delta_{S,T}\rho_S(h)$ for some $h \in G$. Now we must check that

$$\rho_S : \frac{1}{|G|} \sum_{g \in G} [S : D_S]\mathrm{tr}_S(g^{-1}h)g \mapsto \rho_S(h) \, .$$

But

$$\frac{1}{|G|} \sum_{g \in G} [S : D_S]\mathrm{tr}_S(g^{-1}h)\rho_S(g) = \frac{1}{|G|} \sum_{s \in G} [S : D_S]\mathrm{tr}_S(s^{-1})\rho_S(hs)$$

$$= \rho_S(h)\rho_S(e_S) \, ,$$

hence, by Theorem 3.3.4,

$$\frac{1}{|G|} \sum_{g \in G} [S : D_S]\mathrm{tr}_S(g^{-1}h)\rho_S(g) = \rho_S(h) \, .$$

$\square$

3.3.2 Canonical Decomposition of a Representation

Let M be a KG-module. By item (3) of Lemma 3.3.3, it is easy to check that

$$M = \bigoplus_{S \in \mathrm{Irr}_K(G)} e_S M \, .$$

Moreover, if X is an irreducible submodule of M, it follows from Theorem 3.3.4 that

$$e_S X = \begin{cases} X & \text{if } X \simeq S, \\ 0 & \text{if } X \not\simeq S. \end{cases}$$

The next proposition follows from what precedes.

Proposition-Definition 3.3.8 *Let M be a KG-module.*

(1) *For each $S \in \mathrm{Irr}_K(G)$, $e_S M$ is the sum of all submodules of M isomorphic to S.*

(2) *If $M = \bigoplus_{i \in I} S_i$ is a decomposition of M into a direct sum of irreducible modules,*

$$e_S M = \bigoplus_{i \in I_S} S_i$$

where $I_S := \{i \in I \mid S_i \simeq S\}$.

(3) *One has*

$$M = \bigoplus_{S \in \mathrm{Irr}_K(G)} e_S M.$$

For $S \in \mathrm{Irr}_K(G)$, $e_S M$ is called the S-isotypic component of M.

Let us give yet another possible description of the S-isotypic component.

Let M be a KG-module and let S be an irreducible KG-module. Since S is a D_S-vector space, $\mathrm{Hom}_{KG}(S, M)$ is a vector space-D_S (a right vector space over D_S), where this structure is defined by

$$(f \cdot d)(s) := f(ds) \quad \text{for all } f \in \mathrm{Hom}_{KG}(S, M), \, d \in D_S, \, s \in S.$$

Definition 3.3.9 Let M be a KG-module and let S be an irreducible KG-module. The *multiplicity space* is the vector space-D_S defined by

$$\mathrm{Mult}_S(M) := \mathrm{Hom}_{KG}(S, M).$$

Proposition 3.3.10 *Let M be a KG-module. For each irreducible KG-module S, endow the multiplicity space $\mathrm{Mult}_S(M)$ with the trivial action of G. Then the map*

$$\begin{cases} \displaystyle\bigoplus_{S \in \mathrm{Irr}_K(G)} \mathrm{Mult}_S(M) \otimes_{D_S} S \longrightarrow M, \\ (\forall \alpha_S \in \mathrm{Mult}_S(M), \, s \in S), \, \alpha_S \otimes s \mapsto \alpha_S(s), \end{cases}$$

is an isomorphism of KG-modules, such that, for each $S \in \mathrm{Irr}_K(G)$, the image of $\mathrm{Mult}_S(M) \otimes_{D_S} S$ is the S-isotypic component of M.

Proof It is clear that the map is well defined and is a morphism of KG-modules. It is also clear that it is surjective since its image contains every submodule of M

which is isomorphic to S. Hence it suffices to prove that both sides have the same dimension over K. By Proposition 3.2.18, we know that the number $m_{S,M}$ such that $M \cong \bigoplus_{S \in \mathrm{Irr}_K(G)} S^{m_{S,M}}$ is

$$m_{S,M} = \frac{\langle \chi_M, \chi_S \rangle}{\langle \chi_S, \chi_S \rangle} = \frac{[\mathrm{Hom}_{KG}(S, M) : K]}{[D_S : K]}.$$

As just written before Definition 3.3.9, $\mathrm{Hom}_{KG}(S, M)$ is naturally endowed with a structure of vector space-D_S. Thus we have

$$\frac{[\mathrm{Hom}_{KG}(S, M) : K]}{[D_S : K]} = [\mathrm{Hom}_{KG}(S, M) : D_S],$$

and so

$$m_{S,M} = [\mathrm{Hom}_{KG}(S, M) : D_S],$$

which ends the proof. $\square$

Exercise 3.3.11 Let G be a finite group and K be a characteristic zero field. Consider the category where

- Objects are families $(V_S)_{S \in \mathrm{Irr}_K(G)}$ where for each S, V_S is a finite dimensional vector space-D_S (a right D_S-vector space),
- A morphism $f : (V_S)_{S \in \mathrm{Irr}_K(G)} \to (V'_S)_{S \in \mathrm{Irr}_K(G)}$ is a collection of linear-D_S maps $f_S : V_S \to V'_S$.

Prove that this category is equivalent to the category $_{KG}\mathbf{mod}$.

3.4   Group Determinant

Richard Dedekind (1831-1916)

The representation theory of finite groups was developed at the end of the XIX-th century by Ferdinand Georg Frobenius in order to study the group determinant, a problem raised by Richard Dedekind (see [Lam98] on the question of the group determinant; for a general history of representation theory, see [Cur99]). Here we give a treatment in modern language of that question.

Definition 3.4.1 Let $(X_g)_{g \in G}$ be a family of indeterminates indexed by the elements of G, a finite group.

The *group determinant* of G, denoted by $\Theta(G)$, is the determinant of the $|G| \times |G|$ matrix whose (g, h)-th entry (g-th line, h-th column) is $X_{gh^{-1}}$.

In other words, $\Theta(G)$ is the determinant of the endomorphism of the group algebra $\mathbb{Z}[(X_g)_{g \in G}]G$ induced by left multiplication by the "generic" element $\sum_{g \in G} X_g g$.

Thus $\Theta(G)$ is a homogenous element of $\mathbb{Z}[(X_g)_{g \in G}]$ of degree $|G|$.

Assume K is a characteristic zero field, which is a splitting field for the group G. In what follows, we set $\mathrm{Irr}(G) := \mathrm{Irr}_K(G)$. The isomorphism (see Theorem 3.3.1)

$$\prod_{S \in \mathrm{Irr}(G)} \rho_S : KG \xrightarrow{\sim} \prod_{S \in \mathrm{Irr}(G)} \mathrm{End}_K(S)$$

induces in an obvious way an isomorphism

$$\prod_{S \in \mathrm{Irr}(G)} \rho_S : KG \xrightarrow{\sim} \prod_{S \in \mathrm{Irr}(G)} \mathrm{Mat}_{[S:K]}(K), \tag{3.4.2}$$

hence an isomorphism

$$\prod_{S \in \mathrm{Irr}(G)} \rho_S : K[(X_g)_{g \in G}]G \xrightarrow{\sim} \prod_{S \in \mathrm{Irr}(G)} \mathrm{Mat}_{[S:K]}(K[(X_g)_{g \in G}]). \tag{3.4.3}$$

Theorem 3.4.4 (1) *We have*

$$\Theta(G) = \prod_{S \in \mathrm{Irr}(G)} \det \rho_S \left(\sum_{g \in G} X_g g \right).$$

We set $\Theta_S(G) := \det \rho_S(\sum_{g \in G} X_g g)$.
(2) *The family* $(\Theta_S(G))_{S \in \mathrm{Irr}(G)}$ *consists of mutually not associated irreducible elements of* $K[(X_g)_{g \in G}]$.

Proof (1) is immediate from (3.4.3). Let us prove (2).

We may view $(X_g)_{g \in G}$ as the dual basis of the basis $(g)_{g \in G}$ of KG, so that the algebra $K[(X_g)_{g \in G}]$ is identified with the symmetric algebra $S((KG)^*)$ – in other words, the algebra of polynomial functions on KG.

We define another family $(X_{i,j}^{(S)})_{1\leq i,j\leq[S:K],S\in\mathrm{Irr}(G)}$ of linear forms on KG as follows: whenever $a \in KG$ and $S \in \mathrm{Irr}(G)$, the matrix $\rho_S(a)$ has as entries $(X_{i,j}^{(S)}(a))_{1\leq i,j\leq[S:K]}$. In other words, we have

$$\Theta_S(G) = \det\left(X_{i,j}^{(S)}\right)_{1\leq i,j\leq[S:K]}. \qquad (*)$$

This family $(X_{i,j}^{(S)})_{1\leq i,j\leq[S:K],S\in\mathrm{Irr}(G)}$ is also a basis of $(KG)^*$. Indeed, it has the right cardinality (since $|G| = \sum_{S\in\mathrm{Irr}(G)}[S:K]^2$), hence it suffices to prove that it is linearly independent. Assume a dependence relation over K, namely $\sum_{S,i,j} \lambda_{i,j}^{(S)} X_{i,j}^{(S)} = 0$. By the isomorphism (3.4.3), we see that, given S_0, i_0, j_0, there exists $a_{i_0,j_0}^{(S_0)} \in K$ such that $X_{i,j}^{(S)}(a_{i_0,j_0}^{(S_0)}) = \delta_{(S,i,j),(S_0,i_0,j_0)}$, and applying the above dependence relation to $a_{i_0,j_0}^{(S_0)}$ proves that $\lambda_{i_0,j_0}^{(S_0)} = 0$.

Thus the family $(X_{i,j}^{(S)})_{1\leq i,j\leq[S:K],S\in\mathrm{Irr}(G)}$ is yet another family of algebraically independent degree one elements of $S((KG)^*)$. By the equality $(*)$ above, the irreducibility of $\Theta_S(G)$ follows from the well known property of irreducibility of the generic determinant (see Exercise 3.7.15).

The fact that the factors are mutually non associated will result from the following lemma.

Lemma 3.4.5 (1) *Viewed as a polynomial (of degree $[S:K]$) in X_1, we have*

$$\Theta_S(G) = X_1^{[S:K]} + \sum_{g\neq 1}\chi_S(g)X_g X_1^{[S:K]-1} + \cdots .$$

(2) $\Theta_S(G) = \Theta_T(G)$ *if and only if $S = T$.*

Proof (1) Since $\rho_S(\sum_{g\in G} X_g g) = X_1\mathrm{Id}_S + \sum_{g\neq 1} X_g\rho_S(g)$, we see that

- X_1 appears only on the diagonal of the matrix $\rho_S(\sum_{g\in G} X_g g)$, hence the terms of degrees $[S:K]$ and $[S:K] - 1$ in X_1 of $\Theta_S(G)$ only come from the product of the diagonal entries of that matrix,
- the i-th diagonal entry of that matrix is $X_1 + \sum_{g\neq 1} X_g X_{i,i}^{(S)}(g)$,

from which we deduce (1).

(2) By (1) we see that $\Theta_S(G)$ determines χ_S, hence S.   □

          □

3.5 Center and Action of Cyclotomic Galois Groups

Throughout this section, K is a characteristic zero field.

3.5.1 About Centers

General case.

For all $S \in \text{Irr}_K(G)$, let us denote by Z_S the center of the division K-algebra D_S. Thus the center of $\text{End}_{D_S}(S)$ is $Z_S \text{Id}_S$, isomorphic to Z_S.

Since $\prod_S \rho_S$ is an isomorphism, it induces an algebra isomorphism

$$\prod_{S \in \text{Irr}_K(G)} \omega_S : ZKG \xrightarrow{\sim} \prod_{S \in \text{Irr}_K(G)} Z_S . \qquad (3.5.1)$$

The image of an element $x \in ZKG$ under ρ_S is the multiplication by $\omega_S(x) \in Z_S$. The computation of traces shows that

$$\omega_S(x) = \frac{\text{tr}_{S/Z_S}(x)}{[S : Z_S]} . \qquad (3.5.2)$$

In particular, since χ_S^0 acts on S like the multiplication by $\dfrac{|G|}{[S : D_S]}$, we see that

$$\omega_S(\chi_S^0) = \frac{|G|}{[S : D_S]} . \qquad (3.5.3)$$

Splitting field case.

Assume now that K is a splitting field for S. Then $Z_S = K$ and

$$\omega_S : ZKG \to K$$

is an algebra morphism. In particular, whenever C is a conjugacy class of G, $SC = \sum_{g \in C} g$, and $g_C \in C$, we have

$$\omega_S(SC) = \frac{|C|}{[S : K]} \chi_S(g_C) . \qquad (3.5.4)$$

Moreover, since in that case $ZKGe_S = K$, we get

Proposition 3.5.5 *Assume that K is a splitting field for G. Then the family $(e_S)_{S \in \text{Irr}_K(G)}$ is a basis of ZKG as a K-vector space.*

Corollary 3.5.6 *Assume that K is splitting for G.*

(1) *The family $(\chi_S)_{S \in \text{Irr}_K(G)}$ is an orthonormal basis of the vector space $\text{CF}(G, \mathbb{Q}_K^G)$.*
(2) $|\text{Irr}_K(G)| = |\text{Cl}(G)|$.

Proof Since the family $(\chi_S)_{S \in \text{Irr}_K(G)}$ is an orthogonal system in the vector space $\text{CF}(G, \mathbb{Q}_K^G)$, (1) follows from Proposition 3.5.5 above and from Lemma 2.3.17, (1). Then (2) follows from Lemma 2.3.17, (2).   □

Exercise 3.5.7 This exercise provides an alternative proof of Corollary 3.5.6 above.

(1) Let $f \in CF(G, \mathbb{Q}_K^G)$, and let $f^0 = \sum_{g \in G} f(g^{-1})g \in ZKG$. Let (S, ρ_S) be an irreducible representation of G. Prove that

$$\rho_S(f^0) = \frac{|G|}{[S:K]} \langle f, \chi_S \rangle \mathrm{Id}_S .$$

(2) Deduce from (1) that if $f \in CF(G, \mathbb{Q}_K^G)$ is orthogonal to all χ_S for $S \in \mathrm{Irr}_K(G)$, then $f = 0$.
(3) Assume that K is a splitting field for G. Prove that the family $(\chi_S)_{S \in \mathrm{Irr}_K(G)}$ is an orthonormal basis of $CF(G, \mathbb{Q}_K^G)$.

3.5.2 Extending and Restricting Scalars

Let L/K be a field extension.

• Whenever M is a KG-module, we may extend the scalars to L by considering the LG-module $L \otimes_K M$. We have

$$\chi_{L \otimes_K M} = \chi_M .$$

• If L/K is finite, whenever N is an LG-module, we may restrict the scalars to K by considering the underlying structure of K-vector space on N, providing the KG-module N_K. We have computed χ_{N_K} before (Lemma 3.2.5 – see also Proposition 3.5.14 below).

Exercise 3.5.8 Let L/K be a finite field extension. For M a KG-module, prove that

$$\chi_{(L \otimes_K M)_K} = [L:K]\chi_M .$$

Extending the scalars of an irreducible module.

Let $S \in \mathrm{Irr}_K(G)$. Let us denote by $\mathrm{Irr}_L(G, S)$ the set of elements $T \in \mathrm{Irr}_L(G)$ such that T is a constituent of $L \otimes_K S$. Thus

$$\mathrm{Irr}_L(G, S) = \{T \in \mathrm{Irr}_L(G) \mid \langle \chi_S, \chi_T \rangle \neq 0\} .$$

Lemma 3.5.9 (1) $\big(\mathrm{Irr}_L(G, S)\big)_{S \in \mathrm{Irr}_K(G)}$ *is a partition of* $\mathrm{Irr}_L(G)$.
(2) *For each* $S \in \mathrm{Irr}_K(G)$ *and* $T \in \mathrm{Irr}_L(G, S)$, $[S:D_S]$ *divides* $[T:D_T]$ *and*

$$\chi_S = \sum_{T \in \mathrm{Irr}_L(G,S)} \frac{[T:D_T]}{[S:D_S]} \chi_T .$$

Proof Since $(\chi_S)_{S \in \mathrm{Irr}_K(G)}$ is an orthogonal system, we see that if $S \neq S'$, $\langle \chi_T, \chi_S \rangle \neq 0$ and $\langle \chi_{T'}, \chi_{S'} \rangle \neq 0$ imply $T \neq T'$.

Moreover, extending scalars of the regular representation KG provides the regular representation LG.

Hence we have

$$\chi_G^{\mathrm{reg}} = \sum_{S \in \mathrm{Irr}_K(G)} [S : D_S] \chi_S = \sum_{T \in \mathrm{Irr}_L(G)} [T : D_T] \chi_T,$$

which implies that for each $S \in \mathrm{Irr}_K(G)$,

$$\chi_S = \sum_{T \in \mathrm{Irr}_L(G,S)} \frac{[T : D_T]}{[S : D_S]} \chi_T,$$

proving the lemma.   □

Remark 3.5.10 We shall give later (Theorem 4.1.5) a much more precise result about the extension of scalars.

Twisting a representation and restricting scalars.

Definition 3.5.11 Let $\sigma : K \xrightarrow{\sim} K'$ be a field isomorphism, and let M be a KG-module. The $K'G$-module $^\sigma M$ is defined as follows.

- As an abelian (additive) group, $^\sigma M = M$, and for all $g \in G$ the action of g is the same.
- The multiplication by scalars is "twisted by σ" in the following sense: for $m \in {}^\sigma M$ and $\lambda \in K'$, we have $\lambda.m = \sigma^{-1}(\lambda)m$.

Let $(e_i)_{1 \leq i \leq r}$ be a basis of M over K. Then it is also a basis of $^\sigma M$ over K'. For $m \in M$, if $m = \sum_{i=1}^r \lambda_i e_i$ is its expression in M, then its expression on that basis as an element of $^a M$ is $m = \sum_{i=1}^r \sigma(\lambda_i).e_i$.

It follows that if $(\rho(g)_{i,j})$ is the matrix of the action of G on M, then the matrix of the action of G on $^\sigma M$ is $(\sigma(\rho(g)_{i,j}))$. In particular,

$$\text{for all } g \in G, \ \chi_{\sigma M}(g) = \sigma(\chi_M(g)).$$

Remark 3.5.12 Here is an alternative definition of $^\sigma M$.

There is a structure of K-vector space on K', defined by $x.x' := \sigma(x)x'$ for all $x \in K$ and $x' \in K'$. Then $^\sigma M = K' \otimes_K M$.

Remark 3.5.13 Let K be a subfield of the field $\mathbb{C}$ of complex numbers which is stable under the complex conjugation $z \mapsto z^*$.

Then, for any KG-module M, its twist by the complex conjugation is nothing but its contragredient (exercise: why?).

Proposition 3.5.14 *Let L/K be a finite extension, let K' be a Galois extension of K containing L, and let $\mathrm{Mor}_K(L, K')$ denote the set of all field morphisms from L into K' which induce the identity on K.*

Let N be an LG-module. Denote by N_K the KG-module defined by N.

(1) *We have*

$$\chi_{L \otimes_K N_K} = \sum_{\sigma \in \mathrm{Mor}_K(L,K')} \sigma \chi_N \,.$$

(2) *Assume L/K Galois, and assume $K' = L$. Let us denote by $\mathrm{Gal}(L/K)_N$ the fixator of the isomorphism class of N in the Galois group $\mathrm{Gal}(L/K)$, i.e.,*

$$\mathrm{Gal}(L/K)_N = \{\sigma \in \mathrm{Gal}(L/K) \mid {}^\sigma N \simeq N\} \,.$$

Then

$$\chi_{L \otimes_K N_K} = |\mathrm{Gal}(L/K)_N| \sum_{\sigma \in \mathrm{Gal}(L/K)/\mathrm{Gal}(L/K)_N} \sigma \chi_N \,,$$

and we have the following isomorphism of LG-modules:

$$L \otimes_K N_K \simeq \Big(\bigoplus_{\sigma \in \mathrm{Gal}(L/K)/\mathrm{Gal}(L/K)_N} {}^\sigma N \Big)^{|\mathrm{Gal}(L/K)_N|} \,.$$

Proof (1) By Lemma 3.2.5, we know that, for all $g \in G$,

$$\chi_{L \otimes_K N_K}(g) = \chi_{N_K}(g) = \mathrm{tr}_{L/K}(\chi_N(g)) \,,$$

and for $\lambda \in L$, the trace of the K-linear map

$$L \to L \,, \quad x \mapsto \lambda x$$

is $\sum_{\sigma \in \mathrm{Mor}_K(L,K')} \sigma(\lambda)$ (see Proposition B.3.1).

(2) Since a representation is characterized by its character, and since $\chi_{{}^\sigma N} = \sigma \chi_N$, it is an immediate consequence of (1). $\qquad\square$

Remark 3.5.15 Assertion (2) may not be trivial to exhibit in practice, and this again shows the power of characters.

As an example of the situation given by the above Proposition 3.5.14, let us consider the very simple case where $K = \mathbb{R}$ and $L = \mathbb{C}$ (or more generally where K is a real field and $L = K(i)$ for $i^2 = -1$), $M = \mathbb{C}$, and G is a subgroup of $\mu_\mathbb{C}$, acting by multiplication on M.

Then $M_\mathbb{R} = \mathbb{R}.1 \oplus \mathbb{R}.i$, and for $g = x + iy \in G$, the action of g on $M_\mathbb{R}$ is given by the matrix $\begin{pmatrix} x & -y \\ y & x \end{pmatrix}$.

Let us set $\mathbf{1} := 1 \otimes_{\mathbb{R}} 1$ and $\mathbf{I} := 1 \otimes_{\mathbb{R}} i$, so that $(\mathbf{1}, \mathbf{I})$ is a basis of $\mathbb{C} \otimes_{\mathbb{R}} M_{\mathbb{R}}$, and the action of g on $\mathbb{C} \otimes_{\mathbb{R}} M_{\mathbb{R}}$ is given, with regards to that basis, by the above matrix.

Let us set $M_+ := \mathbb{C}(\mathbf{1} + i\mathbf{I})$ and $M_- := \mathbb{C}(\mathbf{1} - i\mathbf{I})$. Then $\mathbb{C} \otimes_{\mathbb{R}} M = M_+ \oplus M_-$, and since

$$\begin{cases} g(\mathbf{1} + i\mathbf{I}) = x\mathbf{1} + y\mathbf{I} - yi\mathbf{1} + ix\mathbf{I} = (x - iy)(\mathbf{1} + i\mathbf{I}) \\ g(\mathbf{1} - i\mathbf{I}) = x\mathbf{1} + y\mathbf{I} + yi\mathbf{1} - ix\mathbf{I} = (x + iy)(\mathbf{1} - i\mathbf{I}), \end{cases}$$

we see that M_+^* is isomorphic to M^* while M_- is isomorphic to M.

Exercise 3.5.16 More generally, let M be any $\mathbb{C}G$-module, let $M_{\mathbb{R}}$ be the underlying real vector space, of dimension $2[M : \mathbb{C}]$, still endowed with the action of G (which is indeed $\mathbb{R}$-linear since it is $\mathbb{C}$-linear).

Let $(e_1, \ldots, e_r)$ be a basis of M. Then $(e_1, \ldots, e_r, ie_1, \ldots, ie_r)$ is a basis of $M_{\mathbb{R}}$. For $\alpha = 1, \ldots, r$, let us set (with a little abuse of notation)

$$e_\alpha := 1 \otimes_{\mathbb{R}} e_\alpha \quad \text{and} \quad f_\alpha := 1 \otimes_{\mathbb{R}} ie_\alpha,$$

and let us consider the subspace M_+ of $\mathbb{C} \otimes_{\mathbb{R}} M_{\mathbb{R}}$ with basis

$$(e_\alpha + if_\alpha)_{1 \leq \alpha \leq r}$$

and the subspace M_- of $\mathbb{C} \otimes_{\mathbb{R}} M_{\mathbb{R}}$ with basis

$$(e_\alpha - if_\alpha)_{1 \leq \alpha \leq r}.$$

Prove that $\mathbb{C} \otimes_{\mathbb{R}} M_{\mathbb{R}} = M_+ \oplus M_-$ and that M_- is a $\mathbb{C}G$-module isomorphic to M, while M_+ is a $\mathbb{C}G$-module isomorphic to M^*.

On numbers of irreducible characters (1).

We shall now compare the number $|\mathrm{Irr}_K(G)|$ of isomorphism classes of irreducible KG-modules and the number $|\mathrm{Irr}_L(G)|$ of isomorphism classes of irreducible LG-modules for a finite Galois extension L/K.

Notation 3.5.17 For K a (characteristic zero) field, we denote by $\mathrm{Cha}_K(G)$ the $\mathbb{Z}$-submodule of the space $\mathrm{CF}(G, \mathbb{Q}_K^G)$ (see Notation 3.2.9) generated by all characters of KG-modules.

The elements of $\mathrm{Cha}_K(G)$ are called the *generalized characters* of G over K.

Lemma 3.5.18 (1) $\mathrm{Cha}_K(G)$ *is the free* $\mathbb{Z}$-*module with basis* $(\chi_S)_{S \in \mathrm{Irr}_K(G)}$, *the family of irreducible* K-*characters of* G.
(2) $\mathrm{Cha}_K(G)$ *is a ring with respect to the pointwise product.*

Proof (1) Whenever M is a KG-module (see Notation 3.2.19)

$$\chi_M = \sum_{S \in \mathrm{Irr}_K(G)} m_{S,M} \chi_S \, ,$$

and this shows that the family $(\chi_S)_{S \in \mathrm{Irr}_K(G)}$ generates $\mathrm{Cha}_K(G)$. It is a free family by item (2) of Proposition 3.2.15.

(2) results from the fact that, for KG-modules M and N, $\chi_M \chi_N = \chi_{M \otimes_K N}$.　□

Lemma 3.5.19 *Let L be a finite Galois extension of K.*

(1) $[L : K]\mathrm{Cha}_K(G) \subset \mathrm{tr}_{L/K}(\mathrm{Cha}_L(G)) \subset \mathrm{Cha}_K(G)$.
(2) *The number $|\mathrm{Irr}_K(G)|$ of isomorphism classes of irreducible KG-modules is equal to the number of orbits of $\mathrm{Irr}_L(G)$ under the Galois group $\mathrm{Gal}(L/K)$.*

Proof (1) Let $S \in \mathrm{Irr}_K(G)$. Then (see Exercise 3.5.8)

$$[L : K]\chi_S = \mathrm{tr}_{L/K}(\chi_{L \otimes_K S}) \, , \text{ and } [L : K]\mathrm{Cha}_K(G) \subset \mathrm{tr}_{L/K}(\mathrm{Cha}_L(G)) \, .$$

Conversely, it follows from Proposition 3.5.14, (2), that

$$\mathrm{tr}_{L/K}(\mathrm{Cha}_L(G)) \subset \mathrm{Cha}_K(G) \, .$$

(2) Both families

$$(\chi_S)_{S \in \mathrm{Irr}_K(G)} \text{ and } (\mathrm{tr}_{L/K}(\chi_T))_{T \in \mathrm{Irr}_L(G)/\mathrm{Gal}(L/K)}$$

(consisting of functions with values in the field $\mathbb{Q}_L^G$) are free over $\mathbb{Q}_L^G$. It follows from (1) that they generate the same $\mathbb{Q}_L^G$-vector space, hence they have the same number of elements.　□

3.5.3　Action of $(\mathbb{Z}/e_G\mathbb{Z})^\times$ and of Galois Groups of Cyclotomic Extensions

Definition and properties.

Denote by e_G the lcm of the orders of the elements of G.

For $n \in (\mathbb{Z}/e_G\mathbb{Z})^\times$ and $g \in G$, the element $g^n \in G$ is well defined. It is clear that the group $G \times (\mathbb{Z}/e_G\mathbb{Z})^\times$ acts on the set G by the formula

$$(x, n).g := x g^n x^{-1} \, ,$$

and in particular for $C \in \mathrm{Cl}(G)$ and $n \in (\mathbb{Z}/e_G\mathbb{Z})^\times$, the set $C^{(n)} := \{g^n \mid g \in C\}$ is again a conjugacy class.

Lemma 3.5.20 *The above action of* $(\mathbb{Z}/e_G\mathbb{Z})^\times$ *on the set* G *induces the following actions.*

(1) *A* $\mathbb{Z}$-*linear action on the group* $\mathrm{CF}(G, \mathbb{Z})$ *of central functions of* G *with values in* $\mathbb{Z}$: *if* f *is a central function on* G, *then* $(^n f)(g) := f(g^n)$.
(2) *A* $\mathbb{Z}$-*linear action on the group* $Z\mathbb{Z}G$: *if* $x = \sum_{g \in G} x_g g \in Z\mathbb{Z}G$, *then* $^n x := \sum_{g \in G} x_g g^n$.

The proof of above lemma is obvious. Notice that, for $C \in \mathrm{Cl}(G)$,

- if γ_C denotes the characteristic function of C, $^n\gamma_C = \gamma_{C^{(n)}}$,
- $^n(SC) = SC^{(n)}$.

Let K be a characteristic zero (commutative) field.

Notation 3.5.21 (1) We recall that we denote by K_G a splitting field of $X^{e_G} - 1$ over K, i.e., (see Notation 3.2.8) $K_G = K(\mu_{e_G}(K))$.
Thus in particular, $\mathbb{Q}_G = \mathbb{Q}(e^{2i\pi/e_G})$.
(2) The Galois group $\mathrm{Gal}(K_G/K)$ can be naturally identified to a subgroup of $\mathrm{Gal}(\mathbb{Q}_G/\mathbb{Q})$, hence to a subgroup $(\mathbb{Z}/e_G\mathbb{Z})_K^\times$ of $(\mathbb{Z}/e_G\mathbb{Z})^\times$:

- for $\sigma \in \mathrm{Gal}(K_G/K)$, we denote by $n(\sigma) \in (\mathbb{Z}/e_G\mathbb{Z})_K^\times$ the element such that, for all e_G-th roots of unity ζ in K_G, we have $\sigma(\zeta) = \zeta^{n(\sigma)}$,
- for $n \in (\mathbb{Z}/e_G\mathbb{Z})_K^\times$, we denote by σ_n the corresponding element of $\mathrm{Gal}(K_G/K)$.

Definition 3.5.22 We call K-*conjugacy classes* of G the orbits of $G \times (\mathbb{Z}/e_G\mathbb{Z})_K^\times$ in its action on G.
Thus two elements $g, h \in G$ belong to the same K-conjugacy class of G if and only if there exists $x \in G$ and $n \in (\mathbb{Z}/e_G\mathbb{Z})_K^\times$ such that $h = xg^n x^{-1}$.

Example 3.5.23 (1) Choose $K = \mathbb{R}$. Then, unless all the elements of G have square 1, we have $\mathbb{R}_G = \mathbb{C}$, and $(\mathbb{Z}/e_G\mathbb{Z})_\mathbb{R}^\times = \{1, -1\}$. In any case, the $\mathbb{R}$-conjugacy class of an element $g \in G$ is the union of the conjugacy class of g and of the conjugacy class of g^{-1}.
(2) Consider the case where $K = \mathbb{Q}$. Then two elements $g, h \in G$ are in the same $\mathbb{Q}$-conjugacy class if and only if the cyclic groups $\langle g \rangle$ and $\langle h \rangle$ they generate are G-conjugate.

The center $Z\mathbb{Z}G$ of the group algebra $\mathbb{Z}G$ is a permutation module for the action of $(\mathbb{Z}/e_G\mathbb{Z})^\times$ defined above Lemma 3.5.20. The following lemma is obvious.

Lemma 3.5.24 *The submodule* $\mathrm{Fix}^{(\mathbb{Z}/e_G\mathbb{Z})_K^\times}(Z\mathbb{Z}G)$ *of* $Z\mathbb{Z}G$ *has for basis the family* (SC) *where* C *runs over the set of* K-*conjugacy classes of* G.
If L *is a field containing* K *as a subfield, the subspace* $\mathrm{Fix}^{(\mathbb{Z}/e_G\mathbb{Z})_K^\times}(ZLG)$ *of* ZLG *has for basis the family* (SC) *where* C *runs over the set of* K-*conjugacy classes of* G.

By Lemma 3.2.32, we may choose a splitting field L for G which is both a finite extension of K_G, and a Galois extension of K. For $\sigma \in \mathrm{Gal}(L/K)$, since K_G/K is Galois, K_G is stable under σ. We still denote by $n(\sigma) \in (\mathbb{Z}/e_G\mathbb{Z})_K^\times$ the element such that $\sigma_{n(\sigma)}$ is the restriction of σ to K_G.

Lemma 3.5.25 *For all $T \in \mathrm{Irr}_L(G)$, $g \in G$, $C \in \mathrm{Cl}(G)$, and $\sigma \in \mathrm{Gal}(L/K)$,*

$$\sigma(\chi_T(g)) = \chi_{\circ T}(g) = \chi_T(g^{n(\sigma)}),$$
$$\sigma(\omega_T(SC)) = \omega_{\circ T}(SC) = \omega_T(SC^{(n(\sigma))}).$$

In particular the action of $\mathrm{Gal}(L/K)$ on the set of characters of LG-modules (as well as on the set of isomorphism classes of LG-modules) is via the epimorphism $\mathrm{Gal}(L/K) \twoheadrightarrow (\mathbb{Z}/e_G\mathbb{Z})_K^{\times}$.

Thus, for $n \in (\mathbb{Z}/e_G\mathbb{Z})_K^{\times}$ and $T \in \mathrm{Irr}_L(G)$, the element $^nT \in (\mathbb{Z}/e_G\mathbb{Z})_K^{\times}$ is defined by the following condition: whenever $\sigma \in \mathrm{Gal}(L/K)$ is such that $n(\sigma) = n$, then

$$^nT = {}^{\sigma}T.$$

Since L is splitting for G, ZLG has for basis the family $(e_T)_{T \in \mathrm{Irr}_L(G)}$, where e_T is the idempotent of ZLG corresponding to T.

Proposition 3.5.26 (1) $e_T \in ZK_GG$.
(2) ZK_GG *has for basis the family* $(e_T)_{T \in \mathrm{Irr}_L(G)}$.
(3) *For* $n \in (\mathbb{Z}/e_G\mathbb{Z})_K^{\times}$, *we have* $^ne_T = e_{^nT}$.

Proof (1) follows from the equality (see Lemma 3.3.7) $e_T = \dfrac{[T:K]}{|G|}\chi_T^0$.
(2) For $x \in ZLG$, we have $x = \sum_{T \in \mathrm{Irr}_L(G)} \omega_T(x)e_T$. To prove (2) it suffices to check that, for all conjugacy class C, $\omega_T(SC) \in K_G$, which is immediate.
(3) follows from (1) and Lemma 3.5.25.   □

Main theorem.

Theorem 3.5.27 (1) *The elements of the group* $(\mathbb{Z}/e_G\mathbb{Z})^{\times}$ *act as ring automorphisms on* $Z\mathbb{Z}G$. *Thus, for all* $n \in (\mathbb{Z}/e_G\mathbb{Z})^{\times}$ *and* $C_1, C_2 \in \mathrm{Cl}(G)$,

$$(SC_1SC_2)^{(n)} = SC_1^{(n)}SC_2^{(n)} .$$

(2) *If K is a characteric zero field, the number* $|\mathrm{Irr}_K(G)|$ *of isomorphism classes of irreducible KG-modules is equal to the number of K-conjugacy classes of G.*

Proof (1) By Proposition 3.5.26, we see that $(\mathbb{Z}/e_G\mathbb{Z})^{\times}$ acts as ring automorphisms on $Z\mathbb{Q}_GG$. Since $Z\mathbb{Z}G$ is stable under $(\mathbb{Z}/e_G\mathbb{Z})^{\times}$, this proves (1).
(2) By Proposition 3.5.26, we know that

$$[\mathrm{Fix}^{(\mathbb{Z}/e_G\mathbb{Z})_K^{\times}}(ZK_GG) : K_G] = |\mathrm{Irr}_L(G)/(\mathbb{Z}/e_G\mathbb{Z})_K^{\times}| .$$

By Lemma 3.5.19, (2), we know that

$$|\mathrm{Irr}_L(G)/(\mathbb{Z}/e_G\mathbb{Z})_K^\times| = |\mathrm{Irr}_K(G)| .$$

By Lemma 3.5.24, we know that $[\mathrm{Fix}^{(\mathbb{Z}/e_G\mathbb{Z})_K^\times}(ZK_GG) : K_G]$ is equal to the number of K-conjugacy classes of G.   □

3.6 More on a Splitting Field and First Applications

3.6.1 Character Tables

The following result is an immediate consequence of Corollary 3.5.6.

Proposition 3.6.1 *Assume that K is splitting for G.*
   For $C \in \mathrm{Cl}(G)$, let us denote by g_C an element of C. Then for all $S \in \mathrm{Irr}_K(G)$ and $C \in \mathrm{Cl}(G)$,

$$\chi_S = \sum_{C \in \mathrm{Cl}(G)} \chi_S(g_C)\gamma_C ,$$

$$\gamma_C = \sum_{S \in \mathrm{Irr}_K(G)} \frac{|C|}{|G|} \chi_S(g_C^{-1})\chi_S .$$

Here we assume K is splitting for G.

The *character table* of a finite group G is the square matrix whose

- columns are indexed by $\mathrm{Cl}(G)$,
- lines are indexed by $\mathrm{Irr}_K(G)$,
- the entry in the box (S, C) is $\chi_S(g_C)$.

Let us equip the C-column with the number $|C|$. Then we say that each entry of the column is *weighted*.

The next fundamental property of character tables follows immediately from Proposition 3.6.1.

Theorem 3.6.2 (1) *The lines (with weighted entries) of the character table of G are orthogonal. More precisely, for all $S, T \in \mathrm{Irr}_K(G)$,*

$$\sum_{C \in \mathrm{Cl}(G)} |C|\chi_S(g_C)\chi_T(g_C)^* = \delta_{S,T}|G| .$$

(2) *The columns of the character table of G are orthogonal. More precisely, for all $C, D \in \mathrm{Cl}(G)$,*

$$\sum_{S \in \mathrm{Irr}_K(G)} \chi_S(g_C)\chi_S(g_D)^* = \delta_{C,D}|G|\frac{1}{|C|} .$$

Notice that a particular case of assertion (2) of the preceding theorem is (see above Proposition 3.2.37, (2))

$$|G| = \sum_{S \in \mathrm{Irr}_K(G)} \chi_S(1)^2 .$$

As examples, let us give the character tables for the groups $\mathfrak{S}_3$, $\mathfrak{S}_4$, $\mathfrak{A}_4$, and the dihedral group of order 8 already mentioned.

Their computation is left as an exercise for the reader. Let us just give some hints.
• The reader may start by writing the part of the character table associated with the characters of degree 1 (Table 3.1).
• We recall that the conjugacy classes of the symmetric group $\mathfrak{S}_n$ are classified by the disjoint cycle decompositions of the elements (Table 3.2).
• It follows from Exercise 2.3.20 that there is a surjective morphism $\mathfrak{S}_4 \twoheadrightarrow \mathfrak{S}_3$. This may be used to determine part of the character table of $\mathfrak{S}_4$ from the character table of $\mathfrak{S}_3$ (using the "inflation" introduce in Sect. 3.2) (Table 3.3).
• The surjective morphism $\mathfrak{S}_4 \twoheadrightarrow \mathfrak{S}_3$ induces a surjective morphism $\mathfrak{A}_4 \twoheadrightarrow C_3$ (where C_3 denotes a cyclic group of order 3), and this provides part of the character table of $\mathfrak{A}_4$. We set $\zeta_3 = \exp(2i\pi/3)$.

We let the reader determine the conjugacy classes of $\mathfrak{A}_4$ (Table 3.4).

Table 3.1 Character table of $\mathfrak{S}_3$

Class$_{|C|}$ →	Irr ↓	1_1	$(12)_3$	$(123)_2$
	1	1	1	1
	ϵ	1	-1	1
	χ	2	0	-1

Table 3.2 Character table of $\mathfrak{S}_4$

Class$_{|C|}$ →	Irr ↓	1_1	$(12)_6$	$(12)(34)_3$	$(123)_8$	$(1234)_6$
	1	1	1	1	1	1
	ϵ	1	-1	1	1	-1
	χ_2	2	0	2	-1	0
	χ_3	3	1	-1	0	-1
	$\epsilon\chi_3$	3	-1	-1	0	1

Table 3.3 Character table of $\mathfrak{A}_4$

	Irr $\downarrow$				
Class$_{\|C\|} \rightarrow$		1_1	$(12)(34)_3$	$(123)_4$	$(132)_4$
	1	1	1	1	1
	χ_1	1	1	ζ_3	ζ_3^*
	χ_1^*	1	1	ζ_3^*	ζ_3
	χ_3	3	-1	0	0

Table 3.4 Character table of the dihedral group of order 8

	Irr $\downarrow$				
Class$_{\|C\|} \rightarrow$		1_1	$((\sigma\tau)_1^2$	$(\sigma)_2$	$(\tau)_2$	$(\sigma\tau)_2$
	ρ_1	1	1	1	1	1
	ρ_τ	1	1	-1	1	-1
	ρ_σ	1	1	1	-1	-1
	$\rho_{\sigma\tau}$	1	1	-1	-1	1
	ρ_2	2	-2	0	0	0

3.6.2 Products of Groups

Let G_1 and G_2 be finite groups, and let M_1, M_2 be respectively a KG_1-module and a KG_2-module. Then $M_1 \otimes_K M_2$ is a $K(G_1 \times G_2)$-module through the law (see for example Proposition 1.2.1):

$$(g_1, g_2).(m_1 \otimes_K m_2) := g_1 m_1 \otimes_K g_2 m_2 .$$

We then have (see Corollary 1.2.12)

$$\chi_{M_1 \otimes_K M_2}(g_1, g_2) = \chi_{M_1}(g_1)\chi_{M_2}(g_2) .$$

Proposition 3.6.3 (1) *Let $S_1 \in \mathrm{Irr}_K(G_1)$ and $S_2 \in \mathrm{Irr}_K(G_2)$. Assume that K is a splitting field for both S_1 and S_2. Then $S_1 \otimes_K S_2$ is an absolutely irreducible $K(G_1 \times G_2)$-module.*

(2) *Assume that K is a splitting field for both G_1 and G_2. Then it is a splitting field for $G_1 \times G_2$, and*

$$\mathrm{Irr}_K(G_1 \times G_2) = \{S_1 \otimes_K S_2 \mid (S_1 \in \mathrm{Irr}_K(G_1))(S_2 \in \mathrm{Irr}_K(G_2))\} .$$

Proof • For $S_1, S_1' \in \mathrm{Irr}_K(G_1)$ and $S_2, S_2' \in \mathrm{Irr}_K(G_2)$,

$$\langle \chi_{S_1 \otimes_K S_2}, \chi_{S'_1 \otimes_K S'_2} \rangle$$

$$= \frac{1}{|G_1 \times G_2|} \sum_{(g_1, g_2) \in G_1 \times G_2} \chi_{S_1}(g_1) \chi_{S_2}(g_2) \chi_{S'_1}(g_1)^* \chi_{S'_2}(g_2)^*$$

$$= \left(\frac{1}{|G_1|} \sum_{g_1 \in G_1} \chi_{S_1}(g_1) \chi_{S'_1}(g_1)^* \right) \left(\frac{1}{|G_2|} \sum_{g_2 \in G_2} \chi_{S_2}(g_2) \chi_{S'_2}(g_2)^* \right)$$

$$= \langle \chi_{S_1}, \chi_{S'_1} \rangle \langle \chi_{S_2}, \chi_{S'_2} \rangle = \delta_{S_1, S'_1} \delta_{S_2, S'_2},$$

which proves that the family $(S_1 \otimes_K S_2)_{\substack{S_1 \in \mathrm{Irr}_K(G_1) \\ S_2 \in \mathrm{Irr}_K(G_2)}}$ consists of distinct absolutely irreducible $K(G_1 \otimes_K G_2)$-modules.

• Since

$$|G_1 \times G_2| = \sum_{\substack{S_1 \in \mathrm{Irr}_K(G_1) \\ S_2 \in \mathrm{Irr}_K(G_2)}} \chi_{S_1}(1)^2 \chi_{S_2}(1)^2 = \sum_{\substack{S_1 \in \mathrm{Irr}_K(G_1) \\ S_2 \in \mathrm{Irr}_K(G_2)}} \chi_{S_1 \otimes_K S_2}(1)^2,$$

that family exhausts $\mathrm{Irr}_K(G_1 \times G_2)$ (by item (2) of Proposition 3.2.37).   □

Remark 3.6.4 The reader is invited to look for a proof not using characters. He might then realize how efficient are the methods of character theory.

3.6.3 Abelian Groups

Proposition 3.6.5 *Let G be a finite group and let K be a splitting field for G. The following assertions are equivalent.*

 (i) *G is abelian,*
 (ii) *for all $S \in \mathrm{Irr}_K(G)$, $[S : K] = 1$.*

Proof The group G is abelian if and only if every conjugacy class has one element, hence if and only if $|G| = |\mathrm{Cl}(G)|$. Since $|\mathrm{Cl}(G)| = |\mathrm{Irr}_K(G)|$, it follows from Corollary 3.5.6 that G is abelian if and only if $|G| = |\mathrm{Irr}_K(G)|$.

On the other hand, since $|G| = \sum_{S \in \mathrm{Irr}_K(G)} [S : K]^2$, we see that $|G| = |\mathrm{Irr}_K(G)|$ if and only if, for all $S \in \mathrm{Irr}_K(G)$, $[S : K] = 1$.

This establishes the proposition.   □

Ⓘ An abelian group may have irreducible representations over K of degree strictly larger than 1 if K is not splitting. For example, the cyclic subgroup $\{\pm 1, \pm i\}$ of $\mathbb{C}^\times$ has an irreducible representation of degree 2 over $\mathbb{Q}$ given by $i \mapsto \begin{pmatrix} 0 & -1 \\ 1 & 0 \end{pmatrix}$.

3.7 Some Arithmetical Properties of Characters

3.7.1 Characters and Integrality

For K a characteristic zero field, we denote by $\mathbb{Z}_K$ the ring of algebraic integers of K. Let G be a finite group. Then we have the following ring extensions.

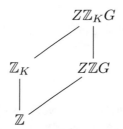

- The elements of the ring $\mathbb{Z}\mathbb{Z}G$ are integral over $\mathbb{Z}$.

  Indeed, it is finitely generated over $\mathbb{Z}$, since $\mathbb{Z}\mathbb{Z}G = \bigoplus_{C \in \mathrm{Cl}(G)} \mathbb{Z}\mathcal{S}C$. Notice that for $C, D \in \mathrm{Cl}(G)$,

$$\mathcal{S}C\mathcal{S}D = \sum_{E \in \mathrm{Cl}(G)} m^E_{C,D}\mathcal{S}E$$

where, for $e \in E$, $m^E_{C,D} = |\{(c,d) \in C \times D \mid cd = e\}|$.
- The elements of the ring $\mathbb{Z}\mathbb{Z}_K G$ are integral over $\mathbb{Z}$.

  Indeed, it consists of all sums and products of elements of $\mathbb{Z}_K$ and of $\mathbb{Z}\mathbb{Z}G$. From now on, *integral element* means "integral over $\mathbb{Z}$".

Proposition 3.7.1 *Let $x \in \mathbb{Z}\mathbb{Z}_K G$. Let S be an irreducible KG-module. We recall that Z_S denotes the center of the field $D_S = \mathrm{End}_{KG}(S)$.*

(1) $\dfrac{\mathrm{tr}_{S/Z_S}(x)}{[S:Z_S]}$ *is an algebraic integer in Z_S.*

(2) $[S:D_S]$ *divides $|G|$.*

(3) *If K is a splitting field for S,*

$$\frac{\chi_S(x)}{\chi_S(1)} \in \mathbb{Z}_K \quad \text{and} \quad [S:K] \text{ divides } |G|.$$

Proof (1) An algebra morphism sends integral elements to integral elements, hence $\omega_S : ZKG \to Z_S$ sends $\mathbb{Z}\mathbb{Z}_K G$ into the set of algebraic integers of Z_S. Then (1) follows from formula (3.5.2).

(2) results then from formula (3.5.3).

(3) is an immediate consequence of (1) and (2). □

The next result makes part of what precedes more precise.

Theorem 3.7.2 *Let H be an abelian normal subgroup of G.*
Whenever S is an absolutely irreducible KG-module,

$$[S : K] \text{ divides } |G : H|.$$

We start by proving a particular case of Theorem 3.7.2.

Lemma 3.7.3 *Let $Z(G)$ denote the center of G. Whenever S is an absolutely irreducible KG-module,*

$$[S : K] \text{ divides } |G : Z(G)|.$$

Proof Let $m \geq 1$ be a natural integer. Then we know by Proposition 3.6.3 that $S^{\otimes m} := S \otimes_K \cdots \otimes_K S$ (m factors) is an absolutely irreducible $K(G^m)$-module. Then we have

$$\omega_{S \otimes_K \cdots \otimes_K S} : \begin{cases} Z(G)^m \to K, \\ (g_1, \ldots, g_m) \mapsto \omega_S(g_1 \cdots g_m). \end{cases}$$

Set

$$Z_m(G) := \{(g_1, \ldots, g_m) \in Z(G)^m \mid g_1 \cdots g_m = 1\}.$$

Since the morphism $Z(G)^m \to Z(G)$, $(g_1, \ldots, g_m) \mapsto g_1 \cdots g_m$, is onto, we have $|Z_m(G)| = |Z(G)|^{m-1}$.

The group $Z_m(G)$ acts trivially on $S^{\otimes m}$, so $S^{\otimes m}$ has a natural structure of absolutely irreducible $K(G^m/Z_m(G))$-module. It then follows from Proposition 3.7.1 that $[S : K]^m$ divides $|G|^m/|Z(G)|^{m-1}$, and so that

$$\text{for all } m \geq 1, \left(\frac{|G : Z(G)|}{[S : K]} \right)^m \in \frac{1}{|Z(G)|} \mathbb{Z}.$$

Thus

$$\mathbb{Z} \left[\frac{|G : Z(G)|}{[S : K]} \right] \subset \frac{1}{|Z(G)|} \mathbb{Z},$$

clearly a finitely generated $\mathbb{Z}$-module, hence $\dfrac{|G : Z(G)|}{[S : K]}$ is integral, hence an integer since $\mathbb{Z}$ is integrally closed in $\mathbb{Q}$.   □

In order to prove Theorem 3.7.2, we need some preliminary general results about normal subgroups.

3.7.1.1 About Representations and Normal Subgroups

Let H be a normal subgroup of G.

Let us first recall that G acts by conjugation on the set of isomorphism classes of KH-modules.

Indeed, for N a KH-module and $g \in G$, the KH-module ${}^gN$ is defined as follows (see Sect. 2.1.2):

- as a K-vector space, ${}^gN = N$,
- for $h \in H$ and $n \in {}^gN$, we set $h.n := (g^{-1}hg)n$.

In particular G acts on $\mathrm{Irr}_K(H)$.

For M a KG-module, we denote by $\mathrm{Res}_H^G M$ the K-vector space M viewed as a KH-module through the inclusion $KH \to KG$ (we shall come back to that notation in Chap. 4).

The proof of the following lemma is easy and left as an exercise to the reader.

Lemma 3.7.4 *Let M be a KG-module and let N be a KH-submodule of $\mathrm{Res}_H^G M$. For $g \in G$,*

(1) *$g(N)$ is also a KH-submodule of $\mathrm{Res}_H^G M$,*
(2) *the map*

$$N \to g(N) , \quad n \mapsto g(n)$$

is an isomorphism ${}^gN \xrightarrow{\sim} g(N)$.

Now let S be an irreducible KG-module. Consider the canonical decomposition of $\mathrm{Res}_H^G S$ into isotypic submodules

$$\mathrm{Res}_H^G S = \bigoplus_{T \in \mathrm{Irr}_K(H)} (\mathrm{Res}_H^G S)_T .$$

An isotypic component $(\mathrm{Res}_H^G S)_T$ is the sum of all KH-submodules of $\mathrm{Res}_H^G S$ which are isomorphic to T. Hence $g((\mathrm{Res}_H^G S)_T)$ is the sum of all KH-submodules of $\mathrm{Res}_H^G S$ which are isomorphic to $g(T)$. By item (2) of Lemma 3.7.4, we have $g(T) \simeq {}^gT$ and so

$$g((\mathrm{Res}_H^G S)_T) = (\mathrm{Res}_H^G S)_{{}^gT} . \tag{3.7.5}$$

Lemma 3.7.6 *Let S be an irreducible KG-module and let T be an irreducible KH-submodule of $\mathrm{Res}_H^G S$. Let G_T denote the subgroup of G comprising those $g \in G$ such that ${}^gT \simeq T$.*

(1) *The isotypic component $(\mathrm{Res}_H^G S)_T$ is an irreducible KG_T-submodule of $\mathrm{Res}_{G_T}^G S$.*
(2) *If $[G/G_T]$ denotes a full set of representatives for left cosets gG_T ($g \in G$), then*

$$S = \bigoplus_{g \in [G/G_T]} g((\mathrm{Res}_H^G S)_T) .$$

Proof The isotypic component $(\mathrm{Res}_H^G S)_T$ is stable under G_T by (3.7.5). We shall prove (2) before proving that it is irreducible as a KG_T-module.

The subspace $\sum_{g \in G} g((\mathrm{Res}_H^G S)_T)$ is stable under G, and since S is irreducible it follows that $S = \sum_{g \in G} g((\mathrm{Res}_H^G S)_T)$.

Since the sum of *distinct* isotypic components is direct, it follows that

$$S = \bigoplus_{g \in [G/G_T]} (\mathrm{Res}_H^G S)_{gT},$$

and since $g((\mathrm{Res}_H^G S)_T) = (\mathrm{Res}_H^G S)_{gT}$, that proves (2).

Now $(\mathrm{Res}_H^G S)_T$ is an irreducible KG_T-module, since for any KG_T-submodule M, $\bigoplus_{g \in [G/G_T]} g(M)$ is a KG-submodule of S.   □

Proof of Theorem 3.7.2. Notice first that we may extend the field K as much as we need. So we assume that K is a splitting field for all the subgroups of G.

Now we consider an abelian normal subgroup H of G and an (absolutely) irreducible KG-module S.

As in Lemma 3.7.6, we consider an irreducible KH-submodule T of $\mathrm{Res}_H^G S$. It follows from item (2) of Lemma 3.7.6 that

$$[S : K] = |G : G_T|[(\mathrm{Res}_H^G S)_T : K].$$

Thus, since $|G : H| = |G : G_T||G_T : H|$, in order to prove that $[S : K]$ divides $|G : H|$, it suffices to prove that $[(\mathrm{Res}_H^G S)_T : K]$ divides $|G_T : H|$.

By our assumption on K, $(\mathrm{Res}_H^G S)_T$ is an absolutely irreducible KG_T-module and T is an absolutely irreducible KH-module.

Then the following lemma (applied to $G = G_T$ and $S = (\mathrm{Res}_H^G S)_T$) ends the proof.

Lemma 3.7.7 *Let H be an abelian normal subgroup of G, let T be an absolutely irreducible KH-module, and let S be an absolutely irreducible KG-module such that $\mathrm{Res}_H^G S$ is T-isotypic. Then $[S : K]$ divides $|G : H|$.*

Proof Let $\rho_S : G \to \mathrm{GL}(S)$ be the structural morphism. Since $|\rho_S(G) : \rho_S(H)|$ divides $|G : H|$, it is enough to prove that $[S : K]$ divides $|\rho_S(G) : \rho_S(H)|$.

Since H is abelian, T is absolutely irreducible, and $\mathrm{Res}_H^G S$ is T-isotypic, it results from Schur's Lemma that $\rho_S(H)$ is contained in the center of $\mathrm{GL}(S)$, hence in the center of $\rho_S(G)$. Then it results from Lemma 3.7.3 that $[S : K]$ divides $|\rho_S(G) : \rho_S(H)|$.   □

□

3.7.2 Applications: Two Theorems of Burnside

Now we use character theory to get purely group theoretic statements.

Burnside's $p^a q^b$-theorem.

Our aim is to prove the following result.

Theorem 3.7.8 (Burnside) *A finite group of order divisible by at most two prime numbers is solvable.*

Let G be a finite group of order divisible by at most two prime numbers. In order to prove this theorem, we argue by induction on $|G|$. We may also assume that $|G|$ is not a prime number (otherwise, G is cyclic).

Assuming this, it is enough to prove that G is not simple. Indeed, if G is not simple there exists a proper nontrivial normal subgroup H of G. Since H and G/H have also orders divisible by at most two prime numbers, the induction hypothesis shows that they are both solvable, hence G is solvable.

Thus it is enough to prove the following proposition.

Proposition 3.7.9 *Let G be a group of order divisible by at most two prime numbers. Then*

- *either G is cyclic of prime order,*
- *or G is not simple.*

Proof The proof relies on three lemmas. The first one is purely arithmetical.

Lemma 3.7.10 *Let $r \geq 1$ and $\zeta_1, \ldots, \zeta_r$ be roots of unity such that $(\zeta_1 + \cdots + \zeta_r)/r$ is an algebraic integer. Then*

- *either $\zeta_1 + \cdots + \zeta_r = 0$,*
- *or $\zeta_1 = \cdots = \zeta_r$.*

Proof of Lemma 3.7.10. Let us set $z := (\zeta_1 + \cdots + \zeta_r)/r$. Let $M_z(X)$ denote its minimal polynomial over $\mathbb{Q}[X]$. Let us choose a Galois extension K' of $\mathbb{Q}$ which contains $\mathbb{Q}(z)$ (one may take $K' = \mathbb{Q}(\zeta)$ where ζ is a root of unity whose order is the lcm of the orders of $\zeta_1, \ldots, \zeta_r$).

We know that

- the roots of $M_z(X)$ are the distinct conjugates of z (that is, images of z under the elements $\sigma \in \mathrm{Gal}(K'/\mathbb{Q})$) – see Proposition B.3.1, (2),
- $M_z(X)$ belongs to $\mathbb{Z}[X]$ since z is an algebraic integer (see Proposition C.2.1).

If $z' = \sigma(z)$, we have $z' = (\zeta_1' + \cdots + \zeta_r')/r$, where $\zeta_j' = \sigma(\zeta_j)$ are roots of unity for $j = 1, \ldots, r$. In particular, we see that $|z'| \leq 1$.

Let us denote by $N(z)$ the product of all the conjugates of z, i.e., up to sign, the constant coefficient of $M_z(X)$. By what precedes, we see that $N(z) \in \mathbb{Z}$ and that $|N(z)| \leq 1$. Hence

- either $N(z) = 0$, in which case at least one of the conjugates of z is zero, hence $z = 0$,
- or $|N(z)| = 1$, i.e., $|\zeta_1 + \cdots + \zeta_r| = |\zeta_1| + \cdots + |\zeta_r|$, which implies $\zeta_1 = \cdots = \zeta_r$.

Lemma 3.7.11 *Let G be a finite group. Assume that S is an absolutely irreducible representation of G and C is a conjugacy class of G such that $[S : K]$ and $|C|$ are relatively prime. Then, whenever $g \in C$,*

- *either g acts on S as a scalar,*
- *or $\chi_S(g) = 0$.*

Proof By Bézout's theorem, there exist $u, v \in \mathbb{Z}$ such that

$$1 = u[S : K] + v|C| .$$

Hence, for $g \in C$,

$$\frac{\chi_S(g)}{[S : K]} = u\chi_S(g) + v\frac{\chi_S(g)|C|}{[S : K]} .$$

By item (3) of Proposition 3.7.1, we know that $\dfrac{\chi_S(g)|C|}{[S : K]}$ is an algebraic integer.
Hence this shows (see Corollary C.1.5) that $\dfrac{\chi_S(g)}{[S : K]}$ is an algebraic integer.

Lemma 3.7.11 follows then from Lemma 3.7.10.   □

Lemma 3.7.12 *Let G be a finite group. Assume that there is a conjugacy class C of G such that $|C|$ is a prime power different from 1. Then G is nonabelian and not simple.*

Proof We choose a splitting field K for G and we set $\mathrm{Irr}(G) := \mathrm{Irr}_K(G)$.

The group G is not commutative since it has a conjugacy class with more than one element.

We assume that C is a conjugacy class of G such that $|C| = p^m$ for some prime number p and some natural integer $m > 0$. Let $g \in C$. Notice that $g \neq 1$ since $|C| > 1$.

Since $g \neq 1$, we have $\chi_G^{\mathrm{reg}}(g) = 0$. Since $\chi_G^{\mathrm{reg}} = \sum_{S\in\mathrm{Irr}(G)}[S : K]\chi_S$, this implies

$$1 + \sum_{S\neq 1}[S : K]\chi_S(g) = 0 .$$

There exists $S \in \mathrm{Irr}(G)$, $S \neq 1$, such that $p \nmid [S : K]$ and $\chi_S(g) \neq 0$. Indeed, if this were not the case, $1/p$ would be an algebraic integer.

Since $|C|$ is a power of p, $|C|$ and $[S : K]$ are relatively prime, and by the preceding lemma it follows that g acts on S as a scalar multiplication.

Since S is not the trivial representation, $\ker(\chi_S)$ is a proper normal subgroup of G. If that subgroup is nontrivial, G is not simple. So let us assume $\ker(\chi_S) = 1$. Since the representation S is faithful, g belongs to the center of G. Hence (since G is non commutative) the center of G is a nontrivial normal subgroup and G is not simple.   □

Now we can prove Proposition 3.7.9.

Assume $|G| = p^a q^b$ where p and q are prime numbers and $a \geq 1$. Let P be a Sylow p-subgroup of G. Thus $|P| = p^a$, and the center $Z(P)$ of P is nontrivial. Let g be a nontrivial element of $Z(P)$. Since P centralizes g, the index in G of the centralizer of g, i.e., the cardinality of the conjugacy class of g, has order prime to p, hence is q^c for some natural integer $c \geq 0$. If $c = 0$, g is central and G is not simple. If $c > 0$, by Lemma 3.7.12, G is not simple. $\qquad\square$

Yet another theorem of Burnside.

Our aim is to prove the following result.

Theorem 3.7.13 (Burnside) *Let χ be a non-linear absolutely irreducible character of a finite group G. There exists $g \in G$ such that $\chi(g) = 0$.*

Proof Let χ be a non-linear absolutely irreducible character of a finite group G. Let us set

$$\Pi_G(\chi) := \prod_{g \in G} \chi(g) .$$

We shall prove that $\Pi_G(\chi) = 0$.

• In order to prove this, we first check that it is enough to prove that $\Pi_G(\chi)$ is an integer.

Indeed, the inequality between the arithmetic and the geometric mean implies

$$\left(\prod_{g \in G} |\chi(g)|^2 \right)^{\frac{1}{|G|}} \leq \frac{1}{|G|} \sum_{g \in G} |\chi(g)|^2 = 1 ,$$

and

(a) $|\Pi_G(\chi)|^2 \leq 1$,
(b) $|\Pi_G(\chi)|^2 = 1$ if and only if, for all $g \in G$, $|\chi(g)|^2 = 1$.

If $\Pi_G(\chi) \in \mathbb{Z}$, (a) shows that $|\Pi_G(\chi)|$ is either 0 or 1. Since $\chi(1) > 1$, (b) implies that $\Pi_G(\chi) = 0$, which proves the theorem.

• Now we prove that $\Pi_G(\chi)$ is an integer, and for that purpose we prove that, for all $\sigma \in \mathrm{Gal}(\mathbb{Q}_G/\mathbb{Q})$, we have $\sigma(\Pi_G(\chi)) = \Pi_G(\chi)$.

We recall that we denote by $\mathcal{CS}(G)$ the set of all cyclic subgroups of G. Moreover, for $A \in \mathcal{CS}(G)$, let us denote by $\mathrm{Gen}(A)$ the set of generators of A. It is clear that

$$\Pi_G(\chi) := \prod_{A \in \mathcal{CS}(G)} \prod_{g \in \mathrm{Gen}(A)} \chi(g) ,$$

so it is enough to prove that, for all cyclic subgroup A of G,

$$\prod_{g \in \text{Gen}(A)} \chi(g)$$

is fixed under the action of $\text{Gal}(\mathbb{Q}_G/\mathbb{Q})$. This follows from the fact that the set $\text{Gen}(A)$ is fixed by the action of $\text{Gal}(\mathbb{Q}_G/\mathbb{Q})$ (see Lemma 3.5.25).   □

The following exercise (a suggestion of Gunter Malle) generalizes Theorem 3.7.13.

Exercise 3.7.14 Let K be a characteristic zero field, let G be a finite group, let S be an irreducible KG-module.

Assume that, given an extension L of K which is a splitting field for G, for all degree one characters $\theta : G \to L^\times$ we have $\langle \chi_S, \theta \rangle = 0$.

Prove that there exists $g \in G$ such that $\chi_S(g) = 0$.

More Exercises on Chap. 3

Exercise 3.7.15 Let k be a field, let $(X_{i,j})_{1 \leq i,j \leq n}$ be a family of n^2 indeterminates. Prove that the determinant of the $n \times n$ matrix with entries $(X_{i,j})_{1 \leq i,j \leq n}$ is an irreducible element of $k[(X_{i,j})_{1 \leq i,j \leq n}]$.

Exercise 3.7.16 Let G be the dihedral group of order 8, generated by elements σ and τ subject to relations $\sigma^2 = \tau^2 = (\sigma\tau)^4 = 1$.

Let C be the subgroup of G generated by $\sigma\tau$. We set $-1 = (\sigma\tau)^2$.

(1) Prove that

$$C = \{1, \sigma\tau, -1, -\sigma\tau\} \quad \text{and} \quad G = \{1, \sigma\tau, -1, -\sigma\tau, \sigma, -\sigma, \tau, -\tau\}.$$

(2) We consider the representation

$$\rho : \begin{cases} G \to GL_2(\mathbb{Q}), \\ \sigma \mapsto \begin{pmatrix} 0 & -1 \\ -1 & 0 \end{pmatrix}, \quad \tau \mapsto \begin{pmatrix} 1 & 0 \\ 0 & -1 \end{pmatrix}. \end{cases}$$

Restricted to C, this morphism defines a $\mathbb{Q}$-representation of C.

(a) Prove that both representations are irreducible.
(b) Prove that $\text{End}_{\mathbb{Q}G}(\mathbb{Q}^2) \cong \mathbb{Q}$.
(c) Compute $\dim \left(\text{End}_{\mathbb{Q}C}(\mathbb{Q}^2) \right)$.
(d) Describe the field $\text{End}_{\mathbb{Q}C}(\mathbb{Q}^2)$.

Exercise 3.7.17 Express the group algebras of $\mathbb{Q}\mathfrak{A}_4$ and $\mathbb{Q}D_8$ as products of matrices algebras.

Exercise 3.7.18 The purpose of this exercise is to prove, without using characters, that $\langle \chi_S, \chi_S \rangle_G$ divides $\langle \chi_M, \chi_S \rangle_G$, that is, $[D_S : K]$ divides $[\text{Hom}_{KG}(M, S) : K]$.

(1) We recall that D_S is a division K-algebra. Find a natural structure of (left) vector space of $\mathrm{Hom}_{KG}(M, S)$ on D_S.

(2) Let D be a division K-algebra, finite dimensional as a K-vector space, and let V be a finite dimensional D-vector space, which then inherits a structure of K-vector space. Check that

$$[V : K] = [V : D][D : K]$$

and conclude.

Exercise 3.7.19 We let G act on itself (from the left) *by conjugation*. We denote by χ_G^{conj} the character associated with the corresponding linear representation (over any field K).

(1) Prove that for all $g \in G$,

$$\chi_G^{\mathrm{conj}}(g) = |C_G(g)|.$$

(2) Use this to prove that the sum of the elements of a line of the character table of G is a nonnegative integer.

Exercise 3.7.20 Let G be a finite group. Let K be a splitting field for G. We set $\mathrm{Irr}(G) := \mathrm{Irr}_K(G)$.

(1) For C, D, E in $\mathrm{Cl}(G)$, we set

$$n_{C,D}^{(E)} = |\{(g, h) \mid (g \in C)(h \in D)(gh \in E\}|.$$

Prove that, for given $g_C \in C$, $g_D \in D$, $g_E \in E$,

$$n_{C,D}^{(E)} = \frac{|C||D||E|}{|G|} \sum_{S \in \mathrm{Irr}(G)} \frac{\chi_S(g_C)\chi_S(g_D)\chi_S(g_E)^*}{\chi_S(1)}.$$

(2) (Frobenius) For $g \in G$, let us denote by $c(g)$ the number of pairs (s, t) $(s, t \in G)$ such that $g = sts^{-1}t^{-1}$. Prove that

$$c(g) = |G| \sum_{S \in \mathrm{Irr}(G)} \frac{\chi_S(g)}{\chi_S(1)}.$$

Exercise 3.7.21 Set $\zeta := \exp(2i\pi/5)$. We denote by $\phi := \dfrac{1 + \sqrt{5}}{2}$ the Golden Ratio.

(1) Set $\psi := \zeta + \zeta^{-1}$. Prove that $\psi^2 + \psi - 1 = 0$.

(2) Deduce from what precedes that

$$\zeta + \zeta^{-1} = \phi - 1 \quad \text{and} \quad \zeta^2 + \zeta^{-2} = -\phi.$$

(3) Let D_{10} be the group generated by two elements σ and τ satisfying

$$\sigma^2 = \tau^2 = 1 \text{ and } (\sigma\tau)^5 = 1.$$

(a) Prove that $D_{10} \simeq C_5 \rtimes C_2$, where C_n denotes a cyclic group of order n.
(b) Prove that D_{10} has 4 conjugacy classes.

(4) Prove that D_{10} has a $\mathbb{Q}(\sqrt{5})$-representation of degree 2

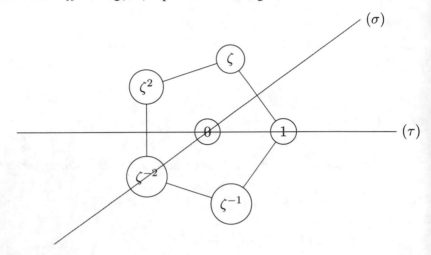

which sends τ onto the symmetry relative to the "x–axis" and σ onto the symmetry relative to the line with angle $\pi/5$ with that x–axis

(5) Prove that the character table of D_{10} is

Class$_{|C|}$ →		1_1	$(\sigma)_5$	$(\sigma\tau)_2$	$((\sigma\tau)^2)_2$
		1	1	1	1
χ_1	1	−1	1	1
χ_2	2	0	$\phi-1$	$-\phi$
χ_2'	2	0	$-\phi$	$\phi-1$

Exercise 3.7.22 The aim is to prove the following property.

Proposition 3.7.23 *Let K be a characteristic zero field and let G be a finite group. Let M be a faithful KG-module, and let $S \in \mathrm{Irr}_K(G)$. Then there exists an integer $n \geq 0$ such that S is a submodule of $M^{\otimes n} = \underbrace{M \otimes \cdots \otimes M}_{n}$.*

HINT. Prove that the formal series $\sum_{n \geq 0} \langle \chi_{M^{\otimes n}}, \chi_S \rangle q^n \in \mathbb{C}[[q]]$ is not zero.

Exercise 3.7.24 Let G be a finite group. We denote by $\det_G : G \to K^\times$ the determinant of the regular representation of G.

(1) Prove that $\det_G(g) = 1$, unless g has even order r such that $|G|/r$ is odd, in which case $\det_G(g) = -1$.
(2) Prove that if G has an element of even order r such that $|G|/r$ is odd, then G has a normal subgroup of index 2.
(3) Assume that $|G| = 2m$ where m is odd. Deduce from what precedes that G has a normal subgroup of order m.

Exercise 3.7.25 We set $\omega := \exp(2\pi i/3)$.
Let us assume that the character table of a finite group G has the following two lines:

	g_1	g_2	g_3	g_4	g_5	g_6	g_7
θ	1	1	1	ω	ω^2	ω	ω^2
χ	2	-2	0	1	1	-1	-1

(1) Determine the complete character table of G. $(!)$ There is a unique solution, do prove it!
(2) Compute $|G|$, as well as the number of elements of each of the 7 conjugacy classes.

Chapter 4
PLAYING with the BASE FIELD

4.1 Analysis of an Irreducible Module

4.1.1 An Example: The Quaternion Group Q_8

The field of quaternions.

Definition 4.1.1 Let K be a subfield of $\mathbb{C}$ which is stable under complex conjugation.

(1) We call *field of quaternions over K* and we denote by $\mathbb{H}(K)$ the subset of $\mathrm{Mat}_2(K)$ defined as

$$\mathbb{H}(K) := \left\{ q(\alpha, \beta) := \begin{pmatrix} \alpha & \beta \\ -\beta^* & \alpha^* \end{pmatrix} \mid \alpha, \beta \in K \right\}.$$

(2) For $q(\alpha, \beta) = \begin{pmatrix} \alpha & \beta \\ -\beta^* & \alpha^* \end{pmatrix} \in \mathbb{H}(K)$, we set

$$q(\alpha, \beta)^* := q(\alpha^*, -\beta) = \begin{pmatrix} \alpha^* & -\beta \\ \beta^* & \alpha \end{pmatrix},$$

and, for $q := q(\alpha, \beta)$,

$$N(q) := qq^*.$$

The following lemma is left as an exercise to the reader (see for example [Bro13], Chap. 1).

© Springer Nature Singapore Pte Ltd. 2017
M. Broué, *On Characters of Finite Groups*, Mathematical Lectures from Peking University, https://doi.org/10.1007/978-981-10-6878-2_4

Lemma 4.1.2 *(1)* $\mathbb{H}(K)$ *is a subring of* $\mathrm{Mat}_2(\mathbb{C})$.
(2) $\mathbb{H}(K)$ *is commutative if and only if* $K \subseteq \mathbb{R}$.
(3) $\mathbb{H}(K)$ *has a natural structure of* $(K \cap \mathbb{R})$*-algebra.*

Remark 4.1.3 Note that $\mathbb{H}(\mathbb{R}) \simeq \mathbb{C}$ and $\mathbb{H}(\mathbb{Q}) \simeq \mathbb{Q}[i]$.

The following properties are straightforward.

(1) For $q := \begin{pmatrix} \alpha & \beta \\ -\beta^* & \alpha^* \end{pmatrix} \in \mathbb{H}(K)$ we have

$$N(q) = |\alpha|^2 + |\beta|^2 .$$

(2) For $q, q' \in \mathbb{H}(K)$, we have

$$(qq')^* = q'^* q^* ,$$
$$N(qq') = N(q)N(q') ,$$

and they imply that $\mathbb{H}(K)$ is a field, 2-dimensional as a K-vector space.

The quaternion group Q_8

From now on, we set $\mathbb{H} := \mathbb{H}(\mathbb{Q}[i])$.
Let us define the following four elements of $\mathbb{H}$:

$$\mathbf{1} := \begin{pmatrix} 1 & 0 \\ 0 & 1 \end{pmatrix}, \ \mathbf{i} := \begin{pmatrix} i & 0 \\ 0 & -i \end{pmatrix}, \ \mathbf{j} := \begin{pmatrix} 0 & 1 \\ -1 & 0 \end{pmatrix}, \ \mathbf{k} := \begin{pmatrix} 0 & i \\ i & 0 \end{pmatrix}.$$

The following properties are left as an exercise to the reader.

(1) If $\alpha = a_1 + i a_2$ and $\beta = b_1 + i b_2$ belong to $\mathbb{Q}[i]$, we have

$$q(\alpha, \beta) = a_1 \mathbf{1} + a_2 \mathbf{i} + b_1 \mathbf{j} + b_2 \mathbf{k} ,$$

hence the family $(\mathbf{1}, \mathbf{i}, \mathbf{j}, \mathbf{k})$ is a basis of $\mathbb{H}$ as a $\mathbb{Q}$-vector space.
(2) These basis elements satisfy the following relations:

$$\mathbf{ij} = \mathbf{k}, \ \mathbf{jk} = \mathbf{i}, \ \mathbf{ki} = \mathbf{j},$$
$$\mathbf{i}^2 = \mathbf{j}^2 = \mathbf{k}^2 = -1.$$

(3) Thus $Q_8 := \{\pm \mathbf{1}, \pm \mathbf{i}, \pm \mathbf{j}, \pm \mathbf{k}\}$ is a noncommutative finite subgroup of order 8 of the multiplicative group of the field $\mathbb{H}$.

Notice that Q_8 has 3 subgroups of order 4, all cyclic, and only 1 (cyclic) subgroup of order 2. Here is the lattice of its subgroups.

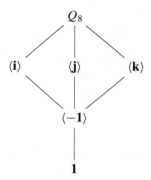

Scalars and group of quaternions

1. An absolutely irreducible module of degree 2 over $\mathbb{Q}[i]$.

The group Q_8 is defined as a subgroup of $GL_2(\mathbb{Q}[i])$, hence is endowed with a faithful representation of degree 2 over $\mathbb{Q}[i]$.

The character χ of that representation is

Class$_{|C|} \rightarrow$	Irr $\downarrow$	$\mathbf{1}_1$	$(-1)_1$	$(\mathbf{i})_2$	$(\mathbf{j})_2$	$(\mathbf{k})_2$
	χ	2	-2	0	0	0

and it is immediate to check that it is absolutely irreducible.

2. An irreducible module of degree 4 over $\mathbb{Q}$.

Letting Q_8 act on $\mathbb{H}$ by left multiplication affords a 4-dimensional $\mathbb{Q}$-representation of Q_8. Matrices of this representation on the basis $(\mathbf{1}, \mathbf{i}, \mathbf{j}, \mathbf{k})$ are easy to compute:

$$\mathbf{i} \mapsto \begin{pmatrix} 0 & -1 & 0 & 0 \\ 1 & 0 & 0 & 0 \\ 0 & 0 & 0 & -1 \\ 0 & 0 & 1 & 0 \end{pmatrix}, \ \mathbf{j} \mapsto \begin{pmatrix} 0 & 0 & -1 & 0 \\ 0 & 0 & 0 & 1 \\ 1 & 0 & 0 & 0 \\ 0 & -1 & 0 & 0 \end{pmatrix}, \ \mathbf{k} \mapsto \begin{pmatrix} 0 & 0 & 0 & -1 \\ 0 & 0 & -1 & 0 \\ 0 & 1 & 0 & 0 \\ 1 & 0 & 0 & 0 \end{pmatrix}$$

Since Q_8 generates (linearly) $\mathbb{H}$, the algebra $\text{End}_{\mathbb{Q}Q_8}(\mathbb{H})$ is also the algebra of $\mathbb{Q}$-endomorphisms of $\mathbb{H}$ which commute with the left multiplications by elements of $\mathbb{H}$.

The proof of the following lemma is left as an exercise to the reader.

Lemma 4.1.4 *Let K be a field and let A be a finite dimensional K-algebra. The algebra $\text{End}_A(A)$ of K-endomorphisms of A which commute with left multiplications by the elements of A is the algebra of right multiplications by elements of A, an algebra isomorphic to A^{op}, the opposite algebra of A.*

By the above lemma, we see that $\text{End}_{\mathbb{Q}Q_8}(\mathbb{H})$ is isomorphic to $\mathbb{H}^{op}$. Since the map $q \mapsto q^*$ is an isomorphism between $\mathbb{H}$ and $\mathbb{H}^{op}$, we have proved that

$$\mathbb{H} \to \mathrm{End}_{\mathbb{Q}Q_8}(\mathbb{H}) \ , \ q \mapsto \left(x \mapsto xq^*\right)$$

is an algebra isomorphism.

It follows (by 3.1.6) that $\mathbb{H}$ is an irreducible $\mathbb{Q}Q_8$-module.

Now the map

$$\mathbb{Q}[i] \to \mathbb{H} \ , \ \lambda \mapsto q(\lambda, 0) = \begin{pmatrix} \lambda & 0 \\ 0 & \lambda^* \end{pmatrix}$$

is a ring morphism and defines a structure of $\mathbb{Q}[i]$-vector space over $\mathbb{H}$. Letting $\mathbb{Q}[i]$ act on $\mathbb{H}$ by right multiplication affords a structure of $\mathbb{Q}[i]Q_8$-module over $\mathbb{H}$. More precisely, for $g \in Q_8$, $\lambda \in \mathbb{Q}[i]$ and $q(\alpha, \beta) \in \mathbb{H}$, let us set

$$\lambda.g : q(\alpha, \beta) \mapsto gq(\alpha, \beta)q(\lambda, 0) = gq(\alpha\lambda, \beta\lambda^*) \,.$$

For that structure of $\mathbb{Q}[i]$-vector space on $\mathbb{H}$, we have

$$\mathbf{i1} = i.\mathbf{1} \quad \text{and} \quad \mathbf{ij} = -i.\mathbf{j} \,,$$

which shows that the matrix of the multiplication by $\mathbf{i}$ on the basis $(\mathbf{1}, \mathbf{j})$ of the $\mathbb{Q}[i]$-vector space $\mathbb{H}$ is $\mathbf{i}$. Similarly, one checks that matrices of multiplications by $\mathbf{j}$ and $\mathbf{k}$ on the basis $(\mathbf{1}, \mathbf{j})$ are respectively $\mathbf{j}$ and $\mathbf{k}$.

In other words, the structure of $\mathbb{Q}[i]Q_8$-module on $\mathbb{H}$ yields the absolutely irreducible module with character χ previously mentioned.

Notice that the character of the $\mathbb{Q}G$-module $\mathbb{H}$ is 2χ.

We shall now show how the general case of arbitrary irreducible modules generalizes the case just described.

4.1.2 Scalars of an Irreducible Module

Let K be, as usual, a field of characteristic zero. Let S be an irreducible KG-module.

- We set

$$D_S := \mathrm{End}_{KG}(S), \ \text{a division } K\text{-algebra},$$
$$Z_S := \text{the center of } D_S, \ \text{a field}.$$

- We know by Theorem 3.3.1 that the image of $\rho_S : KG \to \mathrm{End}_K(S)$ is $\mathrm{End}_{D_S}(S)$.
- We denote by L a maximal commutative subfield of D_S.

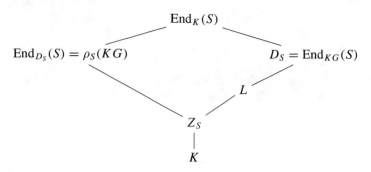

Thus S is naturally endowed with

- a structure of irreducible KG-module (by definition), whose character is denoted by χ_S,
- a structure of irreducible $Z_S G$-module, denoted $S^{(Z_S)}$, whose character is denoted by $\chi_S^{(Z_S)}$,
- a structure of irreducible LG-module, denoted $S^{(L)}$, whose character is denoted by $\chi_S^{(L)}$.

Theorem 4.1.5 *With the above notation,*

(1) *the degree $m_S := [L : Z_S]$ does not depend on the choice of L, and*

$$[D_S : Z_S] = m_S^2 ,$$

(2) *the LG-module $S^{(L)}$ is absolutely irreducible, of degree*

$$\chi_S^{(L)}(1) = [S : D_S]m_S ,$$

(3) $Z_S = K\big[(\chi_S^{(L)}(g))_{g\in G}\big]$, *and in particular Z_S is a Galois extension of K,*

(4) *the $Z_S G$-module $S^{(Z_S)}$ is irreducible with character*

$$\chi_S^{(Z_S)} = m_S \chi_S^{(L)} ,$$

and

$$\chi_S = \mathrm{tr}_{Z_S/K} \cdot \chi_S^{(Z_S)} = \sum_{\sigma\in\mathrm{Gal}(Z_S/K)} \sigma\chi_S^{(Z_S)} = m_S \sum_{\sigma\in\mathrm{Gal}(Z_S/K)} \sigma\chi_S^{(L)} .$$

Remark 4.1.6 It can be proved in general that the dimension of any field over its center is a square (see [Bou12, Sect. 14]).

Proof of Theorem 4.1.5 Since $\mathrm{End}_{LG}(S^{(L)})$ is the set of K-endomorphisms of S which commute both with the actions of all $g \in G$ and of L, it is clear that

$$\mathrm{End}_{LG}(S^{(L)}) = D_S \cap \mathrm{End}_L(S) .$$

Since L is maximal as a commutative subfield of D_S, this implies

$$\mathrm{End}_{LG}(S^{(L)}) = L \,,$$

and $S^{(L)}$ is absolutely irreducible (see Sect. 3.2.28), proving the first part of (2).
The morphism $\omega_S : ZKG \to Z_S$ (see Sect. 3.5.1) is onto, so we have

$$Z_S = K[(\omega_S(\mathcal{S}C))_{C \in \mathrm{Cl}(G)}] \,. \qquad \text{(zs)}$$

By definition, for $C \in \mathrm{Cl}(G)$, we have

$$\omega_S(\mathcal{S}C)\mathrm{Id}_S = \rho_S(\mathcal{S}C) = \rho_{S^{(L)}}(\mathcal{S}C) \,,$$

and so

$$\omega_S(\mathcal{S}C) = \omega_{S^{(L)}}(\mathcal{S}C) \,.$$

By formula 3.5.4, it follows that

$$\omega_{S^{(L)}}(\mathcal{S}C) = \frac{|C|}{[S^{(L)} : L]}\chi_S^{(L)}(g_C) \,,$$

hence

$$\omega_S(\mathcal{S}C) = \frac{|C|}{[S^{(L)} : L]}\chi_S^{(L)}(g_C) \,,$$

and by formula (zs) above we see that the first part of (3) is proved. Thus Z_S is
contained in $K(\mu_{e_G}(K))$, a Galois extension of K with an abelian Galois group,
hence Z_S/K is a Galois extension, and (3) is proved.
    By what precedes, we see that in particular $\chi_S^{(L)}$ takes its values in Z_S, from which
it follows (see for Example 3.2.4) that

$$\chi_S^{(Z_S)} = [L : Z_S]\chi_S^{(L)} = m_S\chi_S^{(L)} \,,$$

which proves the last part of (2), as well as the first part of (4). Since by (3) Z_S/K
is a Galois extension, the whole of (4) follows.
    The above equality also implies

$$\langle \chi_S^{(Z_S)}, \chi_S^{(Z_S)} \rangle = [L : Z_S]^2 \,.$$

But since $\mathrm{End}_{Z_SG}(S^{(Z_S)})$ is the set of K-endomorphisms of S which commute both
with the actions of all $g \in G$ and of Z_S, it is clear that

$$\mathrm{End}_{Z_SG}(S^{(Z_S)}) = D_S \cap \mathrm{End}_{Z_S}(S) = D_S \,.$$

Thus

$$\langle \chi_S^{(Z_S)}, \chi_S^{(Z_S)} \rangle = [D_S : Z_S],$$

proving that

$$[D_S : Z_S] = [L : Z_S]^2,$$

which proves (1) and establishes Theorem 4.1.5. □

4.1.3 Schur Indices

Now we put things "upside down" as far as rationality questions are concerned: instead of starting from a small field and going up to a splitting field, we start with a splitting field and we get down to smaller fields.

For a slightly different approach the reader may refer to [Isa76], Chap. 10.

Let $\mathcal{K}$ be a (commutative) field and let T be an absolutely irreducible $\mathcal{K}G$-module. Let K be a subfield of $\mathcal{K}$. We recall (see Lemma 3.5.9) that there exists a unique (up to isomorphism) irreducible KG-module S such that $\langle \chi_S, \chi_T \rangle \neq 0$.

As in the previous Sect. 4.1.2 we introduce the commuting algebra D_S of S, its center Z_S, and a maximal commutative subfield L of D_S.

Since T is an absolutely irreducible $\mathcal{K}G$-module, we may as well enlarge the field $\mathcal{K}$ (without changing the character χ_T) so that $\mathcal{K}$ contains L.

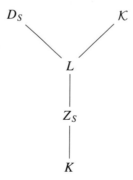

Then (see above Theorem 4.1.5), $S^{(L)}$ is an absolutely irreducible LG-module, and $\mathcal{K} \otimes_L S^{(L)} \simeq T$ and $\chi_{S^{(L)}} = \chi_T$.

The following proposition is essentially a reformulation of Theorem 4.1.5.

Proposition–Definition 4.1.7 *We use notation just introduced above for T, K, S, L.*

(1) If $m_S := [L : Z_S]$, then $[D_S : Z_S] = m_S^2$.

(2) $Z_S = K[(\chi_T(g))_{g \in G}]$ – which we denote by $K(\chi_T)$.

(3) $\chi_S = m_S \sum_{\sigma \in \mathrm{Gal}(Z_S/K)} {}^\sigma \chi_T$.

The integer m_S depends only on T and K. It is denoted $m_K(T)$ and called the Schur index of T relative to K.

Exercise 4.1.8   Compare the above Proposition 4.1.7 with Lemma 3.5.9.

Exercise 4.1.9   Use item (3) of Proposition 3.5.14 to give an alternative proof of the above Proposition 4.1.7.

We shall use the following particular case (and immediate consequence) of the preceding proposition.

Corollary 4.1.10   *Let L be a field and let T be an absolutely irreducible LG-module. Let $K := \mathbb{Q}(\chi_T)$ be the subfield of L generated by $\{\chi_T(g) \mid g \in G\}$, and let S be the unique irreducible KG-module such that $\langle \chi_S, \chi_T \rangle \neq 0$. Let D_S be the endomorphism algebra of the KG-module S.*

(1) $\chi_S = m_K(T)\chi_T$.

(2) $m_K(T)^2 = [D_S : K]$.

(3) *The following assertions are equivalent.*

   (i) $\chi_T = \chi_S$.
   (ii) *T is rational over K.*
   (iii) $D_S = K$.
   (iv) $m_K(T) = 1$.

Let us state (and prove) an omnibus proposition giving many properties of the Schur indices.

Proposition 4.1.11   *Let T be an absolutely irreducible LG-module and let K be a subfield of L.*

(1) $m_K(T) = m_{K(\chi_T)}(T)$.

(2) *$m_K(T)$ divides $[T : L]$.*

(3) *For $n \in \mathbb{N}$ and $n \geq 1$, then $n\chi_T$ is the character of a $K(\chi_T)G$-module if and only if $m_K(T)$ divides n.*

(4) *For $n \in \mathbb{N}$ and $n \geq 1$, then $n\chi_T$ is the character of an irreducible $K(\chi_T)G$-module if and only if $n = m_K(T)$.*

(5) *If K is a subfield of K' and K' is a subfield of L, then*

   (a) *$m_{K'}(T)$ divides $m_K(T)$,*
   (b) *$m_K(T)$ divides $[K' : K]m_{K'}(T)$.*

(6) *Whenever M is a KG-module, then $m_K(T)$ divides $\langle \chi_M, \chi_T \rangle$.*

Proof (1) As above, we use notation and results from Sect. 4.1.2, and denote by S the KG-module defined by restricting the scalars on T from L to K. Thus

$$T = S^{(L)}, \ Z_S = K(\chi_T).$$

The $K(\chi_T)G$-module defined by restricting the scalars on T from L to $K(\chi_T)$ is $S^{(Z_S)}$, and so we see that

$$m_{S^{(Z_S)}} = m_S, \text{ hence } m_K(T) = m_{K(\chi_T)}(T).$$

(2) By Theorem 4.1.5,(2), we see that

$$[T : L] = [S : D_S] m_K(T),$$

which proves (2).

(3) and (4) With the above notation (see Theorem 4.1.5), we have

$$\begin{cases} \chi_S^{(L)} = m_K(T)\chi_T, \\ \chi_S = \displaystyle\sum_{\sigma \in \mathrm{Gal}(K[\chi_T]/K)} \sigma\chi_S^{(L)} = m_K(T) \sum_{\sigma \in \mathrm{Gal}(K[\chi_T]/K)} \sigma\chi_T, \end{cases}$$

and this makes assertions (3) and (4) obvious.

Notice that $\langle \chi_S, \chi_T \rangle = m_K(T)$.

(5)(a) The function $m_K(T)\chi_T$ is the character of a $K(\chi_T)G$-module, hence (since we may extend the scalars from K to K'), it is the character of a $K'(\chi_T)G$-module. By (3), this proves that $m_{K'}(T)$ divides $m_K(T)$.

(5)(b) By (3), the function $m_{K'}(T)\chi_T$ is the character of a $K'(\chi_T)G$-module. The character of that module takes its values in $K'(\chi_T)G$, hence if we view this module as a $K(\chi_T)G$-module its character is $[K'(\chi_T) : K(\chi_T)]m_{K'}(T)\chi_T$. Now by (3) again, we see that $m_K(T)$ divides $[K'(\chi_T) : K(\chi_T)]m_{K'}(T)$, hence divides $[K' : K]m_{K'}(T)$.

(6) If M is a KG-module, its character χ_M is a linear combination with coefficients in $\mathbb{N}$ of characters of irreducible KG-modules. Let us denote by n_S its coefficient on S. Then we have

$$\langle \chi_M, \chi_T \rangle = n_S \langle \chi_S, \chi_T \rangle = n_S m_K(T).$$

$\square$

⚠ **Attention** ⚠

Let L/K be a field extension, and let M be an LG-module.

(1) Suppose that χ_M takes it values in K. Then M need not be rational over K. This is the case, for example, for $K = \mathbb{Q}$, $G = Q_8$, and M the irreducible $\mathbb{Q}(i)Q_8$-module of dimension 2.

(2) But assume now that χ_M is a linear combination with *integral coefficients* of characters of irreducible KG-modules. Then M is rational over K (see below Proposition 4.2.5).

4.2   Complements on Rationality

4.2.1   Group of Characters and Rationality

Generalities

For K a (characteristic zero) field, we recall (see Sect. 3.5.17) that we denote by $\text{Cha}_K(G)$ the subgroup of the space $\text{CF}(G, \mathbb{Q}_K^G)$ (see Sect. 3.2.9) generated by all characters of KG-modules, and (see Lemma 3.5.18) that $\text{Cha}_K(G)$ is the free $\mathbb{Z}$-module with basis the family $(\chi_S)_{S \in \text{Irr}_K(G)}$ of irreducible K-characters of G.

Notation 4.2.1 For L a finite extension of K, let $\text{Cha}_L(G, K)$ denote the subgroup of $\text{Cha}_L(G)$ comprising elements which take their values in K.

Proposition 4.2.2 *For L a finite extension of K, we have the following inclusions*

$$[L : K]\text{Cha}_L(G, K) \subseteq \text{tr}_{L/K}\text{Cha}_L(G) \subseteq \text{Cha}_K(G) \subseteq \text{Cha}_L(G, K).$$

Proof For the convenience of the reader, we repeat some arguments which may have been used before.

- Proof of $\text{Cha}_K(G) \subseteq \text{Cha}_L(G, K)$:if M is a KG-module, then its character χ_M is also the character of the LG-module $L \otimes_K M$, which belongs to $\text{Cha}_L(G, K)$.
- Proof of $\text{tr}_{L/K}\text{Cha}_L(G) \subseteq \text{Cha}_K(G)$:If N is an LG-module with character χ_N, then viewing (by restriction of scalars) N as a KG-module denoted N_K, we have $\text{tr}_{L/K}(\chi_N) = \chi_{N_K}$.
- Proof of $[L : K]\text{Cha}_L(G, K) \subseteq \text{tr}_{L/K}\text{Cha}_L(G)$:If $\chi = \sum a_N\chi_N$ is a linear combination with integral coefficients of characters of LG-modules and if χ takes its values in K, then $[L : K] \chi = \text{tr}_{L/K}\chi = \text{tr}_{L/K}(\sum a_N\chi_N)$.   □

Remark 4.2.3 Each of the inclusions stated in the previous proposition may be strict. We leave as an exercise to the reader to give corresponding examples.

The following statement follows from Proposition 4.2.2 and from Sect. 3.5.18.

Corollary 4.2.4 *The free abelian groups*

$$\text{tr}_{L/K}\text{Cha}_L(G) , \ \text{Cha}_K(G) , \ \text{Cha}_L(G, K),$$

have all the same rank, namely $|\text{Irr}_K(G)|$.

Characterization of rational modules.

Proposition 4.2.5 *Let L be a field extension of K. For X an LG-module, the following assertions are equivalent.*

(i) X is rational over K,
(ii) $\chi_X \in \mathrm{Cha}_K(G)$.

Proof It is clear that (i)$\Rightarrow$(ii). Let us prove (ii)$\Rightarrow$(i). So assume $\chi_X \in \mathrm{Cha}_K(G)$, i.e.,

$$\chi_X = \sum_{S \in \mathrm{Irr}_K(G)} n_{S,X} \chi_S \quad \text{where } n_{S,X} \in \mathbb{Z}.$$

It is enough to prove that $n_{S,X} \in \mathbb{N}$. Indeed, this implies then that

$$X \simeq \bigoplus_{S \in \mathrm{Irr}_K(G)} L \otimes_K S^{n_{S,X}} \simeq L \otimes_K \left(\bigoplus_{S \in \mathrm{Irr}_K(G)} S^{n_{S,X}} \right),$$

proving that X is rational over K.

By the orthogonality relations (see Sect. 3.2.15),

$$n_{S,X} = \frac{\langle \chi_X, \chi_S \rangle}{\langle \chi_S, \chi_S \rangle},$$

and it suffices to prove that $\langle \chi_X, \chi_S \rangle \geq 0$. Since

$$L \otimes_K X \simeq \bigoplus_{T \in \mathrm{Irr}_L(G)} T^{m_{T, L \otimes_K X}}, \quad \text{hence } \chi_X = \sum_{T \in \mathrm{Irr}_L(G)} m_{T, L \otimes_K X} \chi_T,$$

$$L \otimes_K S \simeq \bigoplus_{T \in \mathrm{Irr}_L(G)} T^{m_{T, L \otimes_K S}}, \quad \text{hence } \chi_S = \sum_{T \in \mathrm{Irr}_L(G)} m_{T, L \otimes_K S} \chi_T,$$

we have

$$\langle \chi_X, \chi_S \rangle = \sum_{T \in \mathrm{Irr}_L(G)} m_{T, L \otimes_K X} m_{T, L \otimes_K S} \langle \chi_T, \chi_T \rangle,$$

which shows indeed that $\langle \chi_X, \chi_S \rangle \geq 0$. $\qquad\square$

More precision with Schur indices.

The following proposition is a consequence of Theorem 4.1.5. We also use notation introduced in Sect. 4.1.3 and Proposition–Definition 4.1.7

Proposition 4.2.6 *Assume that L is both a Galois extension of K and a splitting field for G.*

(1) *The family*

$$\left(\mathrm{tr}_{K(\chi_T)/K}(\chi_T)\right)_{T\in\mathrm{Irr}_L(G)/\mathrm{Gal}(L/K)} = ((1/m_S)\chi_S)_{S\in\mathrm{Irr}_K(G)}$$

is a basis of $\mathrm{Cha}_L(G, K)$.

(2) *The family*

$$\left(m_K(T)\mathrm{tr}_{K(\chi_T)/K}(\chi_T)\right)_{T\in\mathrm{Irr}_L(G)/\mathrm{Gal}(L/K)} = (\chi_S)_{S\in\mathrm{Irr}_K(G)}$$

is a basis of $\mathrm{Cha}_K(G)$.

Proof Left as an exercise to the reader.   □

We can then give the following precision to Proposition 4.2.5.

Proposition 4.2.7 *Assume that L is both a Galois extension of K and a splitting field for G. Let X be an LG-module. The following assertions are equivalent.*

(i) X *is rational over* K,
(ii) $\chi_X \in \mathrm{Cha}_K(G)$,
(iii) *For all* $T \in \mathrm{Irr}_L(G)$, $m_K(T)$ *divides* $\langle \chi_X, \chi_T \rangle$.

Proof The equivalence of (i) and (ii) is Proposition 4.2.5. The equivalence of (ii) and (iii) follows from item (2) of Proposition 4.2.6.   □

Complement: on characters of $\mathbb{Q}G$ *-modules.*

Proposition 4.2.8 *Let M and N be $\mathbb{Q}G$-modules. The following assertions are equivalent.*

(i) $M \simeq N$.
(ii) *Whenever A is a cyclic subgroup of G, then*

$$[\mathrm{Fix}^A(M) : \mathbb{Q}] = [\mathrm{Fix}^A(N) : \mathbb{Q}].$$

Proof We only need to prove (ii)⇒(i).

Since χ_M is stable under the action of $(\mathbb{Z}/e_G\mathbb{Z})^\times$, its value on an element g of a cyclic subgroup A of G depends only on the order of g.

By considering the class function $\chi := \chi_M - \chi_N$, we see that in order to prove Proposition 4.2.8 it is enough to prove the following lemma.

Lemma 4.2.9 *Let χ be a class function on G whose value on an element g of a cyclic subgroup A of G depends only on the order of g.*

Assume that, whenever A is a cyclic subgroup of G, $\langle \chi, 1_A \rangle_A = 0$. Then $\chi = 0$.

Proof It is equivalent to prove that whenever A is a cyclic subgroup of G, then $\mathrm{Res}_A^G \chi = 0$. So we may assume (which we do now) that G is cyclic. We argue by induction on $|G|$. Thus $\chi(g) = 0$ whenever g belongs to a proper subgroup of G, which implies $\langle \chi, 1_G \rangle_G = \dfrac{1}{|G|} \varphi(|G|) \chi(g)$ where g denotes a generator of G, and so $\chi(g) = 0$ and $\chi = 0$. $\square$

$\square$

4.2.2 Reflections and Rationality

Definition 4.2.10 Let K be a (characteristic zero) field, and let V be a finite dimensional K-vector space. A *pseudo-reflection* (or simpler, *reflection*) on V is a finite order element $s \in \mathrm{GL}(V)$ such that its space of fixed points $\ker(s - \mathrm{Id}_V)$ is a hyperplane, called its *reflecting hyperplane*.

If s is a reflection on V, then its spectrum is $(\zeta, 1, \dots, 1)$ where ζ is a root of unity. The order of s is equal to the order of ζ, and $\det(s) = \zeta$.

Proposition 4.2.11 *Let K be a (characteristic zero) field, and let T be an absolutely irreducible KG-module. Assume that the image of G in $\mathrm{GL}(T)$ contains a reflection. Then T is rational on the subfield $\mathbb{Q}(\chi_T)$ generated by the values of the character of T.*

Proof We apply Corollary 4.1.10 and refer to notation of Sect. 4.1.2. It suffices to prove that $m_K(T) = 1$. Let S be the unique irreducible $\mathbb{Q}(\chi_T)G$-module such that $\langle \chi_T, \chi_S \rangle \neq 0$, let L be a maximal commutative subfield of D_S. The LG-module $S^{(L)}$ is absolutely irreducible, hence (up to enlarging K so that it contains L), we have $K \otimes_L S^{(L)} \simeq T$.

Thus there is an element $s \in G$ which acts as a reflection on $S^{(L)}$. But its space of fixed points is naturally a D_S-vector space, hence has codimension over L divisible by $[D_S : L] = m_K(T)$, which proves that $m_K(T) = 1$. $\square$

The *finite groups generated by reflections* play an important role in several parts of mathematics.

Let us give some examples of finite subgroups of $\mathrm{GL}_r(\overline{\mathbb{Q}})$ generated by reflections. Let d, e and r be positive integers.

- Let $D_r(de)$ be the set of diagonal complex matrices with diagonal entries in the group μ_{de} of all de–th roots of unity.
- The d–th power of the determinant defines a surjective morphism

$$\det{}^d : D_r(de) \twoheadrightarrow \mu_e .$$

Let $A(de, e, r)$ be the kernel of the above morphism. In particular we have $|A(de, e, r)| = (de)^r/e$.

- Identifying the symmetric group $\mathfrak{S}_r$ with the usual $r \times r$ permutation matrices, we define

$$G(de, e, r) := A(de, e, r) \rtimes \mathfrak{S}_r . \tag{4.2.12}$$

We have $|G(de, e, r)| = (de)^r r!/e$, and $G(de, e, r)$ is the group of all monomial $r \times r$ matrices, with entries in μ_{de}, and product of all non-zero entries in μ_d.

For any positive integer m, let us set $\zeta_m := \exp(2\pi i/m)$.

Exercise 4.2.13 (1) Check that if $d > 1$, $G(d, 1, r)$ is generated by the reflection

$$s_d := \begin{pmatrix} \zeta_d & 0 & 0 & \cdots & 0 \\ 0 & 1 & 0 & \cdots & 0 \\ 0 & 0 & 1 & \cdots & 0 \\ \vdots & \vdots & \vdots & \ddots & \vdots \\ 0 & 0 & \cdots & \cdots & 1 \end{pmatrix}$$

and by the reflections associated with the permutation matrices defined by the transpositions.

(2) Check that if $d, e > 1$ $G(de, e, r)$ is generated by the reflection s_d, by the reflections associated with the permutation matrices defined by the transpositions, and by the order 2 reflection

$$t_{de} :- \begin{pmatrix} 0 & \zeta_{de} & 0 & \cdots & 0 \\ \zeta_{de}^{-1} & 0 & 0 & \cdots & 0 \\ 0 & 0 & 1 & \cdots & 0 \\ \vdots & \vdots & \vdots & \ddots & \vdots \\ 0 & 0 & \cdots & \cdots & 1 \end{pmatrix}$$

(3) What happens for $d = 1$ and $e > 1$?

Exercise 4.2.14 (1) Check that, for all (d, e, r) such that $(d, e, r) \neq (1, 2, 2)$ and $(d, e) \neq (1, 1)$, $G(de, e, r)$ acts irreducibly on $\mathbb{C}^r$.

(2) What happens for $d = e = 1$?

(3) What happens for $de = e = r = 2$?

Shephard and Todd [ST54] have classified all the irreducible finite subgroups of $GL_r(\mathbb{C})$ generated by reflections up to conjugacy. More will be said about these groups in Chap. 6 below.

- For $r > 8$, there are only the groups $G(de, e, r)$ with $de > 1$ and the symmetric group $\mathfrak{S}_{r+1}$ in its natural representation of dimension r.
- For $r \leq 8$, there are moreover 34 exceptional irreducible complex reflection groups, denoted $G_4, G_5, \ldots, G_{37}$.

The rank 2 groups are related to the finite subgroups of $SL_2(\mathbb{C})$ (the *binary polyhedral groups*) – see for example [Ben93].

Let G be one of the above Shephard–Todd groups, acting irreducibly as a group generated by reflections on a complex vector space V.

By Proposition 4.2.11, we know that this representation is rational over the field $\mathbb{Q}(\chi_V) = \mathbb{Q}((\chi_V(g))_{g \in G})$.

A case by case analysis of all absolutely irreducible representations of all Shephard–Todd groups allows to prove the following theorem (see [Bes97] or [Ben76]).

Theorem 4.2.15 *Let G be one of the Shephard–Todd groups, acting irreducibly as a group generated by reflections on a complex vector space V. Then all irreducible representations of G are rational on the field $\mathbb{Q}(\chi_V)$.*

Thus for these groups there is a *smallest field* of rationality for all representations. (!) This is not the case in general: see below Exercise 4.2.22. No general (classification free proof) of that result is known so far.

Let G be one of the Shephard–Todd groups, and let $\mathbb{Q}_G$ denote its smallest field of rationality.

- If $\mathbb{Q}_G \subseteq \mathbb{R}$, the group G is a (finite) *Coxeter group*.
- If $\mathbb{Q}_G = \mathbb{Q}$, the group G is a finite *Weyl group*.

Example 4.2.16 • $G(e, e, 2)$ is the dihedral group of order $2e$.
- $G(d, 1, r)$ is isomorphic to the wreath product $\mu_d \wr \mathfrak{S}_r$. For $d = 2$, it is isomorphic to the Weyl group of type B_r (or C_r).
- $G(2, 2, r)$ is isomorphic to the Weyl group of type D_r.

4.2.3 Questions of Rationality over $\mathbb{R}$

Generalities

Proposition 4.2.17 (1) *Let K be a subfield of the field $\mathbb{R}$ of real numbers and let $i \in \mathbb{C}$ such that $i^2 = -1$. Let V be a $K(i)G$-module. The following assertions are equivalent.*

(i) *V is rational over K, i.e., there exists a KG-submodule V_0 of V such that $V = V_0 \oplus iV_0$.*

(ii) *There exists a KG-automorphism $\sigma : V \to V$ such that $\sigma^2 = \mathrm{Id}_V$ and $\sigma(\lambda v) = \lambda^* \sigma(v)$ for all $\lambda \in K(i)$ and $v \in V$.*

(2) *Assume $K = \mathbb{R}$. Then (i) and (ii) are equivalent to*

(iii) *There exists a nondegenerate symmetric bilinear form on V invariant under G.*

Proof (1) (i)⇒(ii) : For $v = v_0 + iv_0' \in V$ (with $v_0, v_0' \in V_0$), define $\sigma(v) :=$ $v_0 - iv_0'$. Then for $\lambda = x + iy$ with $x, y \in \mathbb{R}$,

$$\sigma(\lambda v) = \sigma((x + iy)(v_0 + iv_0')) = \sigma(xv_0 - yv_0' + i(xv_0' + yv_0))$$
$$= xv_0 - yv_0' - i(xv_0' + yv_0) = (x - iy)(v_0 - iv_0') = \lambda^* \sigma(v).$$

(ii)⇒(i) : Since $\sigma^2 = \mathrm{Id}_V$, we have $V = V_1 \oplus V_{-1}$ where $V_1 := \ker(\sigma - 1)$ and $V_{-1} := \ker(\sigma + 1)$. Since σ is antilinear, multiplication by i induces KG-linear isomorphisms $V_1 \xrightarrow{\sim} V_{-1}$ and $V_{-1} \xrightarrow{\sim} V_1$.

(2) (i)⇒(iii) : Choose any positive definite symmetric bilinear form s on V_0. Then the form

$$s_G(v, v') := \frac{1}{|G|} \sum_{g \in G} s(gv, gv') \quad (\text{for } v, v' \in V)$$

is still positive definite symmetric bilinear, and is invariant under G. Then the form $\widehat{s}_G$ defined on $V = \mathbb{C} \otimes_{\mathbb{R}} V$ by

$$\widehat{s}_G(\lambda \otimes_K v, \lambda' \otimes_K v') := \lambda \lambda' s_G(v, v')$$

is indeed a nondegenerate symmetric bilinear form on V invariant under G.

(iii)⇒(ii) : Let s be a nondegenerate symmetric bilinear form on V invariant under G. Choose any positive definite hermitian form h on V. Then the form

$$h_G(v, v') := \frac{1}{|G|} \sum_{g \in G} h(gv, gv') \quad (\text{for } v, v' \in V)$$

is still positive definite hermitian, and is invariant under G. It is immediate to check that there exists a unique antilinear automorphism φ of the $\mathbb{C}G$-module V such that

$$s(v, v') = h_G(\varphi(v), v')^* = h_G(v', \varphi(v)).$$

The map φ^2 is a linear automorphism of V. It is hermitian relative to h_G, since, by definition of φ,

$$h_G(\varphi^2(v), v') = s(\varphi(v), v')^* = s(v', \varphi(v))^* = h_G(\varphi(v'), \varphi(v))$$
$$= h_G(\varphi(v), \varphi(v'))^* = h_G(\varphi^2(v'), v)^* = h_G(v, \varphi^2(v')).$$

Moreover, since $h_G(\varphi^2(v), v) = h_G(\varphi(v), \varphi(v))$, φ^2 is positive definite relative to h_G (that is, the form $h_G(\varphi^2(\cdot), \cdot)$ is positive definite).

Thus there exists a positive definite hermitian (relative to h_G) automorphism ψ of V such that $\psi^2 = \varphi^2$ and such that ψ commutes with φ^2. Indeed, there exists an orthonormal basis for h_G where the matrix of φ^2 is diagonal with eigenvalues (λ) where $\lambda \in \mathbb{R}$, $\lambda > 0$. Then we choose ψ to be the automorphism whose matrix on the same basis is the diagonal matrix with eigenvalues $(\sqrt{\lambda})$ (where $(\sqrt{\lambda})$ denotes the positive square root of λ).

Now let us set $\sigma := \varphi\psi^{-1}$. Then $\sigma^2 = \mathrm{Id}_V$ and σ is antilinear.   □

The three types of irreducible complex representations

Exercise 4.2.18 (*Frobenius Theorem*) Let $\mathbb{D}$ be a division $\mathbb{R}$-algebra, of finite dimension over $\mathbb{R}$. Prove that one of the following holds:

(R) $\mathbb{D}$ is isomorphic to $\mathbb{R}$,
(C) $\mathbb{D}$ is isomorphic to $\mathbb{C}$,
(H) $\mathbb{D}$ is isomorphic to $\mathbb{H}(\mathbb{R})$ (see Definition 4.1.1 for the notation).

HINT: *Assume* $[\mathbb{D} : \mathbb{R}] \geq 2$.

(1) *Prove that there is an element* $i \in \mathbb{D}$ *such that* $i^2 = -1$. *Prove that* $\mathbb{R}(i) \cong \mathbb{C}$.
(2) *Consider* $\mathbb{D}_+ := \{x \in \mathbb{D} \mid xi = ix\}$ *and* $\mathbb{D}_- := \{x \in \mathbb{D} \mid xi = -ix\}$. *Prove that* $\mathbb{D} = \mathbb{D}_+ \oplus \mathbb{D}_-$ *and that* $\mathbb{D}_+ = \mathbb{C}$.
(3) *If* $\mathbb{D}_- \neq \{0\}$, *prove that there exists* $j \in \mathbb{D}_-$ *such that* $j^2 = -1$ *and that* $D_- = \mathbb{R}(i)j$
(4) *Prove that* $\mathbb{D} = \mathbb{R} \oplus \mathbb{R}i \oplus \mathbb{R}j \oplus \mathbb{R}ij$.

The next theorem distinguishes three different possibilities for an irreducible $\mathbb{C}G$-module relative to $\mathbb{R}$.

Theorem 4.2.19 *Let S be an irreducible $\mathbb{C}G$-module, with character χ. There are three mutually exclusive possibilities:*

(C) *The character χ takes at least one nonreal value.*
     Then S is not rational over $\mathbb{R}$, and by restriction of scalars S defines an $\mathbb{R}G$-module $S_\mathbb{R}$, with character $\chi_\mathbb{R} = \chi + \chi^*$.

   (a) *The module $S_\mathbb{R}$ is irreducible.*
   (b) *The commuting field $D_{S_\mathbb{R}}$ is isomorphic to $\mathbb{C}$.*
   (c) *The Schur index of S relative to $\mathbb{R}$ is 1.*

(R) *The $\mathbb{C}G$-module S is rational over $\mathbb{R}$.*
     Let S_0 be an $\mathbb{R}G$-module such that $S \simeq \mathbb{C} \otimes_\mathbb{R} S_0$.

   (a) $\chi_{S_0} = \chi$, *and χ takes its values in $\mathbb{R}$.*
   (b) *The commuting field D_{S_0} is equal to $\mathbb{R}$.*
   (c) *The Schur index of S relative to $\mathbb{R}$ is 1.*

(H) *The character χ takes its values in $\mathbb{R}$ but the $\mathbb{C}G$-module S is not rational over $\mathbb{R}$.*

   (a) *By restriction of scalars S defines an $\mathbb{R}G$-module $S_\mathbb{R}$, with character $\chi_\mathbb{R} := 2\chi$.*

(b) *The commuting field $D_{S_\mathbb{R}}$ is isomorphic to the field of quaternions $\mathbb{H}$.*

(c) *The Schur index of S relative to $\mathbb{R}$ is 2.*

Proof The situations (C), (R) and (H) are clearly the only possibilities, and are mutually exclusive.

Assume (C). Since χ is not real, it results from 3.2.5 that $\chi_\mathbb{R} = \chi + \chi^*$, hence that $\langle \chi_\mathbb{R}, \chi_\mathbb{R} \rangle = 2$, which implies both that $S_\mathbb{R}$ is an irreducible $\mathbb{R}$-module and that its commuting field $D_{S_\mathbb{R}}$ has dimension 2 over $\mathbb{R}$. Since that field contains $\mathbb{C}$, this proves that it is equal to $\mathbb{C}$. Thus (a) and (b) hold. Assertion (c) follows again from the equality $\chi_\mathbb{R} = \chi + \chi^*$, by item (3) 4.1.7.

If (R) holds, assertions (a), (b), (c) are obvious.

Assume (H). Then, by 3.2.5, we see that $\chi_\mathbb{R} = 2\chi$. By item (3) of 4.1.7, we see that the Schur index is 2. Hence assertions (a) and (c) hold. Since now $\langle \chi_\mathbb{R}, \chi_\mathbb{R} \rangle = 4$, assertion (b) follows from Frobenius theorem (see Exercise 4.2.18).   □

The next theorem uses the Frobenius–Schur indicator defined in 3.2.29.

Theorem 4.2.20 *Let S be an irreducible $\mathbb{C}G$-module with character χ_S. In each item (1), (2), (3), assertions (i), (ii) and (iii) are equivalent.*

(1) (i) *(R) holds.*

(ii) *There is a symmetric nondegenerate bilinear form on S invariant under G.*

(iii) $\nu_2(\chi_S) = 1$.

(2) (i) *(C) holds.*

(ii) *There is no nonzero bilinear form on S invariant under G.*

(iii) $\nu_2(\chi_S) = 0$.

(3) (i) *(H) holds.*

(ii) *There is an alternating bilinear form on S invariant under G.*

(iii) $\nu_2(\chi_S) = -1$.

Proof This is an immediate consequence of Theorem 3.2.30 and of Proposition 4.2.17.   □

More Exercises on Chap. 4

Exercise 4.2.21 In this exercise we use notation defined in Sect. 4.1.1.

(1) Prove that

    (a) the 8 elements $(\mathbf{1} \pm \mathbf{i} \pm \mathbf{j} \pm \mathbf{k})/2$ all have multiplicative order 6,

    (b) the 8 elements $(-\mathbf{1} \pm \mathbf{i} \pm \mathbf{j} \pm \mathbf{k})/2$ all have multiplicative order 3.

(2) Prove that the set

$$\{\pm\mathbf{1}, \pm\mathbf{i}, \pm\mathbf{j} \pm \mathbf{k}\} \cup \{(\mathbf{1} \pm \mathbf{i} \pm \mathbf{j} \pm \mathbf{k})/2\} \cup \{(-\mathbf{1} \pm \mathbf{i} \pm \mathbf{j} \pm \mathbf{k})/2\}$$

    is a group G (under multiplication), of order 24, isomorphic to the semi-direct product $Q_8 \rtimes C_3$ (where C_3 denotes a cyclic subgroup of order 3).

(3) Prove that $G \simeq \mathrm{SL}_2(3)$ (the group of (2×2)-matrices with entries in the finite prime field $\mathbb{F}_3$ and determinant 1).

(4) The group G is given as a subgroup of $\mathrm{GL}(\mathbb{Q}[i])$. Is the corresponding $\mathbb{Q}[i]G$-module absolutely irreducible?

Exercise 4.2.22 Let $K := \mathbb{Q}(i, \sqrt{2})$. Let $Q_8 = \{\pm\mathbf{1}, \pm\mathbf{i}, \pm\mathbf{j}, \pm\mathbf{k}\}$ be the quaternion group of order 8, viewed as a subgroup of $\mathbb{H}(K)^{\times}$. Thus Q_8 is a subgroup of $\mathrm{GL}_2(K)$, and $S := K^2$ may be viewed as a $K Q_8$-module.

(1) Prove that S is rational over $\mathbb{Q}(i)$, and defines an absolutely irreducible $\mathbb{Q}(i)Q_8$-module.

(2) Prove that S is rational over $\mathbb{Q}(i\sqrt{2})$, and defines an absolutely irreducible $\mathbb{Q}(i\sqrt{2})Q_8$-module.

    HINT. One can use the matrices

$$\mathbf{i}' := \begin{pmatrix} i\sqrt{2} & 1 \\ 1 & -i\sqrt{2} \end{pmatrix}, \quad \mathbf{j}' := \begin{pmatrix} 0 & -1 \\ 1 & 0 \end{pmatrix}, \quad \mathbf{k}' := \begin{pmatrix} 1 & -i\sqrt{2} \\ -i\sqrt{2} & -1 \end{pmatrix}.$$

(3) Check that both $\mathbb{Q}(i)$ and $\mathbb{Q}(i\sqrt{2})$ are minimal splitting subfields of $\mathbb{Q}(i, \sqrt{2})$, and that they are non isomorphic.

Exercise 4.2.23 Let K be a (characteristic zero) field, let S be an irreducible KG-module, let $D_S := \mathrm{End}_{KG}S$ be its commuting algebra, let L be a maximal commutative subfield of D_S, and let $T = S^{(L)}$ be the absolutely irreducible LG-module naturally defined on S (see Theorem 4.1.5). Prove that

$$\chi_S = \mathrm{tr}_{L/K}(\chi_T).$$

Exercise 4.2.24

- Let D be the dihedral group of order 8.
  Thus, D is isomorphic to the group of isometries of a square. It is also isomorphic to a Sylow 2-subgroup of the symmetric group $\mathfrak{S}_4$, and to the non-commutative semi-direct product of a cyclic group of order 4 by the group of order 2.
- Let Q be the quaternion group of order 8.

(1) DESCRIPTION OF D AND Q.

    (a) Prove that D may be described as follows

$$D = \{1, -1, \sigma, -\sigma, \tau, -\tau, \sigma\tau, -\sigma\tau\}$$

    where
- σ and $-\sigma$ are two elements of order 4,
- $\sigma^2 = -1$,
- $-1, \tau, -\tau, \sigma\tau, -\sigma\tau$ have order 2.

    Give a complete set of representatives of the conjugacy classes of D.

    (b) Let us set

$$Q = \{1, -1, I, -I, J, -J, IJ, -IJ\}.$$

    Give a complete set of representatives of the conjugacy classes of Q.

(2) Prove that D and Q are not isomorphic.

(3) Compute the character tables of both D and Q.

(4) Deduce that

$$\mathbb{C}D \simeq \mathbb{C}Q.$$

(5) Prove that

$$\mathbb{R}D \not\simeq \mathbb{R}Q.$$

Exercise 4.2.25 Let G be a finite group. For C a conjugacy class of G, we set $C^* := \{g^{-1} \mid g \in C\}$. We say that C is *real* if $C = C^*$.

(1) Prove that the number of irreducible complex characters of G which take only real values equals the number of real conjugacy classes of G.

(2) Prove that if G had odd order, every non trivial absolutely irreducible character of G takes at least one nonreal value.

Chapter 5
Induction, Restriction

5.1 On Any Field

Throughout this section, G denotes a finite group, H a subgroup of G, and k is a commutative field. Notice that here we make no assumption about the characteristic of k.

5.1.1 Restriction

Let X be a kG-module, *i.e.*, a k-vector space endowed with a group morphism $G \rightarrow \mathrm{GL}(X)$. Restricting that morphism to H gives a structure of kH-module on X, which we denote by $\mathrm{Res}_H^G X$.

A morphism of kG-modules $\alpha : X \rightarrow X'$ is in particular a morphism of kH-modules $\mathrm{Res}_H^G(\alpha) : \mathrm{Res}_H^G X \rightarrow \mathrm{Res}_H^G X'$.

It is easy to check that Res_H^G is a functor $_{kG}\mathbf{mod} \rightarrow {}_{kH}\mathbf{mod}$.

Remark 5.1.1 Let us notice a possible (more complicated) definition of Res_H^G.

Let X be a kG-module.

View kG as a kH-module-kG through left multiplication by kH and right multiplication by kG. Thus $kG \otimes_{kG} X$ is a kH-module *via* the rule $h.(a \otimes_{kG} x) := ha \otimes_{kG} x$. Then the morphisms

$$kG \otimes_{kG} X \rightarrow \mathrm{Res}_H^G X \,,\ g \otimes_{kG} x \mapsto gx$$
$$\mathrm{Res}_H^G X \rightarrow kG \otimes_{kG} X \,,\ x \mapsto 1 \otimes_{kG} x$$

are inverse isomorphisms (of kH-modules).

© Springer Nature Singapore Pte Ltd. 2017
M. Broué, *On Characters of Finite Groups*, Mathematical Lectures from Peking University, https://doi.org/10.1007/978-981-10-6878-2_5

5.1.2   Induction

Definition.
   View kG as a kG-module-kH through left multiplication by kG and right multiplication by kH. For a kH-module Y, we define

$$\operatorname{Ind}_H^G Y := kG \otimes_{kH} X .$$

Thus $\operatorname{Ind}_H^G Y$ is a kG-module *via* the rule $g.(a \otimes_{kH} x) := ga \otimes_{kH} x$.
   If $\beta : Y \to Y'$ is a kH-morphism, then the map defined by

$$\operatorname{Ind}_H^G \beta : \begin{cases} \operatorname{Ind}_H^G Y \to \operatorname{Ind}_H^G Y' , \\ g \otimes_{kH} y \mapsto g \otimes_{kH} \beta(y) \end{cases}$$

is a morphism of kG-modules.
   Then it is easy to check that $\operatorname{Ind}_H^G : {}_{kH}\mathbf{mod} \to {}_{kg}\mathbf{mod}$ is a functor.
   Induction, like restriction, is transitive, in the following sense.

Proposition 5.1.2 *Assume that L is a subgroup of H which is a subgroup of G. Then*

$$\operatorname{Res}_L^G = \operatorname{Res}_L^H \cdot \operatorname{Res}_H^G \quad and \quad \operatorname{Ind}_L^G = \operatorname{Ind}_H^G \cdot \operatorname{Ind}_L^H .$$

Proof The transitivity of restriction is obvious. For the induction, the reader may check that, for all kL-module Z, the maps

$$\operatorname{Ind}_L^G Z \to \operatorname{Ind}_H^G \cdot \operatorname{Ind}_L^H Z , \ g \otimes_{kL} z \mapsto g \otimes_{kH} 1 \otimes_{kL} z ,$$
$$\operatorname{Ind}_H^G \cdot \operatorname{Ind}_L^H Z \to \operatorname{Ind}_L^G Z , \ g \otimes_{kH} h \otimes_{kL} z \mapsto gh \otimes_{kL} z ,$$

are well defined, and inverse isomorphisms of kG-modules.   □

   The following property is essential.

Lemma 5.1.3 *Let H be a subgroup of G, let Y be a kH-module and let $s \in G$. Then the map*

$$\operatorname{Ind}_H^G Y \to \operatorname{Ind}_{{}^sH}^G {}^sY , \ g \otimes_{kH} y \mapsto gs^{-1} \otimes_{k^sH} y ,$$

induces an isomorphism of kG-modules.

Proof It is enough to check that this map is well defined, *i.e.*, that the map

$$kG \times Y \to \operatorname{Ind}_{{}^sH}^G {}^sY , \ (g, y) \mapsto gs^{-1} \otimes_{k^sH} y \ \text{ for } g \in G \text{ and } y \in Y ,$$

is such that, for all $h \in H$, the images of (gh, y) and (g, hy) are equal. This is the case, since

$$ghs^{-1} \otimes_{k^sH} y = gs^{-1}shs^{-1} \otimes_{k^sH} y = gs^{-1} \otimes_{k^sH} hy\,.$$

$\square$

Let us notice that we have already met an important example of induced module (the proof of the next lemma is left to the reader).

Lemma 5.1.4 *Let 1_H be k viewed as endowed with the trivial action of H and let $k(G/H)$ denote the kG-module defined by the transitive G-set G/H. Then the map*

$$\operatorname{Ind}_H^G 1_H \to k(G/H)\,, \quad g \otimes_{kH} 1 \mapsto gH$$

is an isomorphism of kG-modules.

Let Y be a kH-module.

For $g \in G$, the subspace $g \otimes_{kH} Y := \{(g \otimes_{kH} y) \mid y \in Y\}$ of $kG \otimes_{kH} Y$ depends only on the coset gH. We have (with obvious notation)

$$kG \otimes_{kH} Y = \bigoplus_{g \in G/H} g \otimes_{kH} Y\,. \tag{5.1.5}$$

Clearly, $1 \otimes_{kH} Y$ is stable under the action of H. The maps

$$\iota_Y : \begin{cases} Y \to kG \otimes_{kH} Y\,, \\ y \mapsto 1 \otimes_{kH} y \end{cases}$$

$$\pi_Y : \begin{cases} kG \otimes_{kH} Y \to Y\,, \\ g \otimes_{kH} y \mapsto \begin{cases} gy & \text{if } g \in H \\ 0 & \text{if } g \notin H \end{cases} \end{cases}$$

are kH-morphisms, inducing inverse isomorphisms between $1 \otimes_{kH} Y$ and Y.

Universal property

Let us denote by ι the collection of all morphisms of kH-modules

$$\iota_Y : Y \to \operatorname{Res}_H^G \operatorname{Ind}_H^G Y \quad \text{for all } kH\text{-modules } Y\,.$$

In the language of categories, ι is called a morphism of functors from the identity functor of $_{kH}\mathbf{mod}$ to the functor $\operatorname{Res}_H^G \operatorname{Ind}_H^G$.

The pair $(\operatorname{Ind}_H^G, \iota)$ is characterized by the following universal property.

Proposition 5.1.6 (Universal property of $(\operatorname{Ind}_H^G, \iota)$) *If (X, f) is a pair where*

- *X is a kG-module,*
- *$f : Y \to \operatorname{Res}_H^G X$ is a kH-morphism,*

there exists a unique kG-morphism $\hat{f} : \operatorname{Ind}_H^G Y \to X$ such that the following diagram is commutative

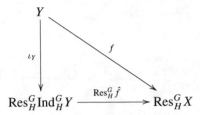

Sketch of proof Indeed, $\hat{f}$ has to satisfy and is defined by

$$\hat{f} : g \otimes_{kH} y \mapsto g f(y) .$$

$\square$

The next corollary expresses the fact that induction is left adjoint to restriction.

Corollary 5.1.7 (Frobenius reciprocity) *The map $f \mapsto \hat{f}$ defines a natural bijection*

$$\mathrm{Hom}_{kH}(Y, \mathrm{Res}_H^G X) \xrightarrow{\sim} \mathrm{Hom}_{kG}(\mathrm{Ind}_H^G Y, X) .$$

As an exercise, the reader may prove the following explicit description of the above isomorphism.

Proposition 5.1.8 *The k-linear maps*

$$\begin{cases} \mathrm{Hom}_{kH}(Y, \mathrm{Res}_H^G X) \to \mathrm{Hom}_{kG}(\mathrm{Ind}_H^G Y, X) , \\ \beta \mapsto \left(\displaystyle\sum_{g \subset G/H} g \otimes_{kH} y_g \mapsto \sum_{g \in G/H} g \beta(y_g) \right) , \end{cases}$$

and

$$\begin{cases} \mathrm{Hom}_{kG}(\mathrm{Ind}_H^G Y, X) \to \mathrm{Hom}_{kH}(Y, \mathrm{Res}_H^G X) , \\ \alpha \mapsto (y \mapsto \alpha(1 \otimes_{kH} y)) , \end{cases}$$

are inverse isomorphisms.

Let us reformulate what precedes in a particular case.

Let X be a kG-module, and let Y be a submodule of $\mathrm{Res}_H^G X$.

Then the subspace $\sum_{g \in G} g(Y)$ of X is G-stable, hence it is a kG-submodule of X. The image of the injection $\iota : Y \hookrightarrow \mathrm{Res}_H^G X$ through the isomorphism described in Corollary 5.1.7 is the kG-morphism

$$\mathrm{Ind}_H^G Y \to X , \quad g \otimes_{kH} y \mapsto g(y) ,$$

whose image is the submodule $\sum_{g \in G} g(Y)$.

Proposition 5.1.9 *Let H be a subgroup of G, let X be a kG-module and let Y be a kH-submodule of $\mathrm{Res}_H^G X$. Then the following properties are equivalent.*

(i) $X = \bigoplus_{g \in G/H} g(Y)$,

(ii) *the map* $\mathrm{Ind}_H^G Y \to X$, $g \otimes_{kH} y \mapsto g(y)$, *is an isomorphism of kG-modules.*

The previous proposition has an immediate application to the dimension of irreducible modules.

Proposition 5.1.10 *Let S be an irreducible kG-module, and let T be a kH-submodule of $\mathrm{Res}_H^G S$.*

(1) There is a surjective morphism $\mathrm{Ind}_H^G T \twoheadrightarrow S$ *such that, for all $t \in T$, $1 \otimes_{kH} t \mapsto t$.*
(2) $[S : k] \leq |G : H|[T : k]$.
(3) The above morphism $\mathrm{Ind}_H^G T \twoheadrightarrow S$ *is an isomorphism if and only if $[S : k] = |G : H|[T : k]$.*

Induction and tensor product.
The following proposition connects induction, restriction, and tensor products.

Proposition 5.1.11 *Let H be a subgroup of G, let Y be a kH-module and let X be a kG-module. Then the k-linear maps defined by*

$$\begin{cases} kG \otimes_{kH} X \otimes_k Y \to X \otimes_k kG \otimes_{kH} Y, \\ g \otimes_{kH} x \otimes_k y \mapsto gx \otimes_k g \otimes_{kH} y, \end{cases}$$

$$\begin{cases} X \otimes_k kG \otimes_{kH} Y \to kG \otimes_{kH} X \otimes_k Y, \\ x \otimes_k g \otimes_{kH} y \mapsto g \otimes_{kH} g^{-1}x \otimes_k y, \end{cases}$$

are inverse isomorphisms of kG-modules

$$\mathrm{Ind}_H^G(\mathrm{Res}_H^G X \otimes_k Y) \longleftrightarrow X \otimes_k \mathrm{Ind}_H^G Y.$$

Remark 5.1.12 One has an analogous definition for induction of H-sets. Indeed, let Y be an H-set. Then H acts on the set $G \times Y$ by

$$h \cdot (g, y) := (gh^{-1}, hy) \quad \text{for all } g \in G, y \in Y, h \in H,$$

and we denote by $G \times_H Y$ the set of orbits of H on $G \times Y$.

For $(g, y) \in G \times Y$, we denote by $g \times_H y$ its orbit. Thus for all $g \in G$, $y \in Y$, $h \in H$, we have

$$gh \times_H y = g \times_H hy,$$

For $g \in G$, the subset $g \times_H Y := \{g \times_H y \mid y \in Y\}$ of $G \times_H Y$ depends only on the coset gH. We have

$$G \times_H Y = \bigsqcup_{g \in G/H} g \times_H Y .$$

We denote by $\operatorname{Ind}_H^G(Y)$

- the set $G \times_H Y$ of orbits of $G \times Y$ under H,
- endowed with the action of G defined by

$$g_1 \cdot (g \times_H y) := (g_1 g \times_H y) .$$

If $\beta : Y \to Y'$ is an H-morphism, then the map

$$\operatorname{Ind}_H^G(\beta) : \begin{cases} \operatorname{Ind}_H^G Y \to \operatorname{Ind}_H^G Y' , \\ g \times_H y \mapsto g \times_H \beta(y) \end{cases}$$

is a morphism of G-sets. Thus $\operatorname{Ind}_H^G : {}_H\mathbf{Set} \to {}_G\mathbf{Set}$ is a functor.

Another definition of induction.
If Y is a kH-module, the set $\operatorname{Hom}_{kH}(\operatorname{Res}_H^G kG, Y)$ is endowed with a natural structure of kG-module, defined by

$$g \cdot \beta(g_1) := \beta(g_1 g^{-1}) \quad \text{for all } g, g_1 \in G, \ \beta \in \operatorname{Hom}_{kH}(\operatorname{Res}_H^G kG, Y) .$$

Since $\operatorname{Res}_H^G kG = \bigoplus_{r \in [H \backslash G]} kHr$, we see that

$$\operatorname{Hom}_{kH}(\operatorname{Res}_H^G kG, Y) = \operatorname{Hom}_{kH}\Big(\bigoplus_{r \in [H \backslash G]} kHr, Y \Big)$$

$$= \prod_{r \in [H \backslash G]} \operatorname{Hom}_{kH}(kHr, Y) .$$

Exercise 5.1.13 Why did I choose to use different signs ($\bigoplus_{r \in [H \backslash G]}$ and $\prod_{r \in [H \backslash G]}$) for analogous constructions ? You might find an answer in [Mac71].

Moreover, let us set

$$\kappa_Y : \begin{cases} Y \to \operatorname{Hom}_{kH}(\operatorname{Res}_H^G kG, Y) , \\ y \mapsto \left(g \mapsto \begin{cases} gy & \text{if } g \in H , \\ 0 & \text{if } g \notin H . \end{cases} \right) \end{cases}$$

Then κ_Y induces an isomorphism of kH-modules

$$\kappa_Y : Y \overset{\sim}{\longrightarrow} \operatorname{Hom}_{kH}(kH, Y) .$$

Proposition 5.1.14 *(1) The pair* $(\operatorname{Hom}_{kH}(\operatorname{Res}_H^G kG, \cdot), \kappa)$ *satisfies the universal property described in Proposition 5.1.6.*

(2) There is a unique isomorphism of kG-modules

$$\sigma : \operatorname{Ind}_H^G Y \xrightarrow{\sim} \operatorname{Hom}_{kH}(\operatorname{Res}_H^G kG, Y)$$

such that the following diagram is commutative:

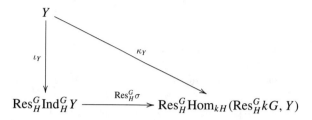

Proof Assertion (2) is a routine and easy consequence of (1). Now we prove (1).
Let (X, f) be a pair where

- X is a kG-module,
- $f : Y \to \operatorname{Res}_H^G X$ is a kH-morphism.

In order to satisfy the universal property, there must exist a unique kG-morphism
$\bar{f} : \operatorname{Hom}_{kH}(\operatorname{Res}_H^G kG, Y) \to X$ such that

$$\operatorname{Res}_H^G \bar{f} \cdot \kappa_Y = f .$$

Since

$$\operatorname{Hom}_{kH}(\operatorname{Res}_H^G kG, Y) = \prod_{r \in [H \backslash G]} \operatorname{Hom}_{kH}(kHr, Y) ,$$

and since $\bar{f}$ must be a kG-morphism, for all $\beta \in \operatorname{Hom}_{kH}(kHr, Y)$ we have $r^{-1} f \in \operatorname{Hom}_{kH}(kH, Y)$ and we must have

$$\bar{f}(\beta) = r \bar{f}(r^{-1}\beta) ,$$

which shows that $\bar{f}$ is determined by its action on $\operatorname{Hom}_{kH}(kH, Y)$. Now since κ_Y
defines an isomorphism $Y \xrightarrow{\sim} \operatorname{Hom}_{kH}(kH, Y)$, $\bar{f}$ is indeed determined by the condition $\operatorname{Res}_H^G \bar{f} \cdot \kappa_Y = f$.

Let us give now a definition of $\bar{f}$. We set

$$\bar{f} : \begin{cases} \operatorname{Hom}_{kH}(kHg, Y) \to X , \\ \beta \mapsto g^{-1} f(\beta(g)) . \end{cases}$$

We leave to the reader the proof that $\bar{f}$ is indeed a kG-morphism. The condition
$\operatorname{Res}_H^G \bar{f} \cdot \kappa_Y = f$ is clear by construction. $\qquad\square$

Proposition 5.1.15 (Induction right adjoint to restriction) *Let X be a kG-module and let Y be a kH-module. The maps*

$$\begin{cases} \mathrm{Hom}_{kH}(\mathrm{Res}_H^G X, Y) \to \mathrm{Hom}_{kG}(X, \mathrm{Hom}_{kH}(\mathrm{Res}_H^G kG, Y)), \\ \beta \mapsto (x \mapsto (g \mapsto \beta(gx))) \end{cases}$$

$$\begin{cases} \mathrm{Hom}_{kG}(X, \mathrm{Hom}_{kH}(\mathrm{Res}_H^G kG, Y)) \to \mathrm{Hom}_{kH}(\mathrm{Res}_H^G X, Y), \\ \alpha \mapsto (x \mapsto \alpha(x)(1)) \end{cases}$$

are mutually inverse isomorphisms.
  Hence they define an isomorphism

$$\mathrm{Hom}_{kH}(\mathrm{Res}_H^G X, Y) \xrightarrow{\sim} \mathrm{Hom}_G(X, \mathrm{Ind}_H^G Y).$$

Proof An immediate verification left to the reader.   □

Remark 5.1.16 In the language of categories, the isomorphism (see Corollary 5.1.7)

$$\mathrm{Hom}_{kH}(Y, \mathrm{Res}_H^G X) \xrightarrow{\sim} \mathrm{Hom}_{kG}(\mathrm{Ind}_H^G Y, X)$$

expresses the fact that induction is left adjoint to restriction, while the above isomorphism

$$\mathrm{Hom}_{kH}(\mathrm{Res}_H^G X, Y) \xrightarrow{\sim} \mathrm{Hom}_G(X, \mathrm{Ind}_H^G Y).$$

expresses the fact that induction is right adjoint to restriction.

Let us describe the above isomorphism

$$\mathrm{Hom}_{kH}(\mathrm{Res}_H^G X, Y) \xrightarrow{\sim} \mathrm{Hom}_{kG}(X, \mathrm{Ind}_H^G Y).$$

Let $\beta : \mathrm{Res}_H^G X \to Y$ be an H-morphism. Let us first notice that, for $x \in X$ and $g \in G$, the element $g \otimes_{kH} \beta(g^{-1}x)$ depends only on the coset gH.

Proposition 5.1.17 *The k-linear maps*

$$\begin{cases} \mathrm{Hom}_{kH}(\mathrm{Res}_H^G X, Y) \longrightarrow \mathrm{Hom}_{kG}(X, \mathrm{Ind}_H^G Y), \\ \beta \mapsto \left(x \mapsto \displaystyle\sum_{g \in G/H} g \otimes_{kH} \beta(g^{-1}x) \right), \end{cases}$$

and

$$\begin{cases} \mathrm{Hom}_{kG}(X, \mathrm{Ind}_H^G Y) \longrightarrow \mathrm{Hom}_{kH}(\mathrm{Res}_H^G X, Y)\,, \\[2mm] \alpha : \left(x \mapsto \displaystyle\sum_{g \in G/H} g \otimes_{kH} y_g \right) \mapsto (x \mapsto y_1)\,, \end{cases}$$

are inverse isomorphisms.

5.1.3 Mackey Formula

Some notation

- Let H and L be subgroups of G. Then the group $L \times H$ acts on G by

$$(l, h) \cdot g := lgh^{-1}\,.$$

We denote by $L \backslash G / H$ the set of orbits, called *double cosets of G modulo* (L, H). For $g \in G$, its double coset is denoted LgH.

- We recall the following notation.

For $g \in G$, we denote by $\mathrm{Inn}(g) : G \to G$ the inner automorphism defined by $\mathrm{Inn}(g) : x \mapsto gxg^{-1}$. We also set ${}^g x := \mathrm{Inn}(g)(x)$, and if H is a subgroup of G, we set ${}^g H := gHg^{-1} = \mathrm{Inn}(g)(H)$.

If Y is a G-set (resp. a kG-module), we denote by $\mathrm{Inn}(g)(Y)$ or by ${}^g Y$ the ${}^g H$-set (resp. $k\,{}^g H$-module) defined on Y by the composition of morphisms

$$ {}^g H \xrightarrow{\;\;\mathrm{Inn}(g^{-1})\;\;} H \longrightarrow \mathfrak{S}(Y)\ (\text{resp. } \mathrm{GL}(Y))\,. $$

Remark 5.1.18 For Y a kH-module, the subset $g \otimes_{kH} Y$ of $\mathrm{Ind}_H^G Y$ is naturally endowed with a structure of $k\,{}^g H$-module.

The map ${}^g Y \to kG \otimes_{kH} Y$, $y \mapsto g \otimes_{kH} y$ is an isomorphism.

The Mackey Theorem

Theorem 5.1.19 (Mackey's formula for modules) *Let H and L be subgroups of G. Let $[L \backslash G / H]$ be a complete set of representatives of the double cosets of G modulo (L, H).*

We have the following equality of functors from ${}_{kH}\mathbf{mod}$ to ${}_{kL}\mathbf{mod}$:

$$\mathrm{Res}_L^G \circ \mathrm{Ind}_H^G = \bigoplus_{r \in [L \backslash G / H]} \mathrm{Ind}_{L \cap {}^r H}^L \circ \mathrm{Res}_{L \cap {}^r H}^{{}^r H} \circ \mathrm{Inn}(r)\,.$$

Proof We leave to the reader to check that this is nothing but an immediate consequence of the following lemma.

Lemma 5.1.20 *For each $r \in [L \backslash G/H]$, let $[L/L \cap {}^r H]$ be a complete set of representatives of cosets of L modulo $L \cap {}^r H$. Then we get a complete set of representatives $[G/H]$ of cosets of G modulo H by the formula*

$$[G/H] := \bigsqcup_{r \in [L \backslash G/H]} [L/L \cap {}^r H] r .$$

Proof The partition into double cosets gives

$$G = \bigcup_{\substack{l \in L \\ r \in [L \backslash G/H]}} lr H .$$

Since, for $l \in L$,

$$(lr H = r H) \Longleftrightarrow (l \in L \cap {}^r H)$$

we get the desired formula.   □

                                                                                            □

Remark 5.1.21 Lemma 5.1.20 provides the same formula as Theorem 5.1.19 for induction and restriction functors in corresponding categories $_H$**Set** and $_L$**Set**.

The following corollary results from Mackey's formula Theorem 5.1.19 and from the adjunction formulas Proposition 5.1.15 and Corollary 5.1.7.

Corollary 5.1.22 *Let H and L be subgroups of G, let Y be a kH-module, let Z be a kL-module. Then*

$$\mathrm{Hom}_{kG}(\mathrm{Ind}_H^G Y, \mathrm{Ind}_L^G Z) = \bigoplus_{r \in [L \backslash G/H]} \mathrm{Hom}_{k(L \cap {}^r H)}(\mathrm{Res}_{L \cap {}^r H}^{{}^r H} {}^r Y, \mathrm{Res}_{L \cap {}^r H}^L Z) .$$

Particular cases
Assume that A and G' are subgroups of G such that $A \subseteq G'$ and $A \lhd G$.
We shall apply Mackey's theorem to the case where $L = A$ and $H = G'$. Then

- $A \backslash G/G' = G/G'$,
- $A \cap {}^g G' = {}^g A \cap {}^g G' = {}^g(A \cap G') = {}^g A = A$,
- $\mathrm{Res}_A^{{}^g G'} \mathrm{Inn}(g) = {}^g \mathrm{Res}_A^{G'}$.

Thus we get the following particular case of the Mackey formula.

Corollary 5.1.23 *Assume that A and G' are subgroups of G such that $A \subseteq G'$ and $A \lhd G$. Then*

$$\mathrm{Res}_A^G \mathrm{Ind}_{G'}^G = \bigoplus_{g \in [G/G']} {}^g \mathrm{Res}_A^{G'} .$$

5.2 Induction and Restriction in Characteristic Zero

From now on, in this paragraph, we denote by K a *characteristic zero* field.
   Let us notice the following immediate consequence of Proposition 5.1.10.

Proposition 5.2.1 *Assume H abelian and K a splitting field for H, and let S be an irreducible KG-module. Then*

$$[S : K] \leq |G : H|.$$

5.2.1 Induction and Normal Subgroups

By Proposition 5.1.9, Lemma 3.7.6 may be reformulated as follows.

Proposition 5.2.2 *Let S be an irreducible KG-module. Let H be a normal subgroup of G. For T an irreducible KH-submodule of $\mathrm{Res}_H^G S$, we denote by $(\mathrm{Res}_H^G S)_T$ the T-isotypic component of $\mathrm{Res}_H^G S$, and by G_T the fixator of T (sometimes called the "inertia group" of T), that is, $G_T := \{g \in G \mid {}^g T \simeq T\}$. Then*

(1) $(\mathrm{Res}_H^G S)_T$ is stable under G_T, and as such inherits a structure of irreducible KG_T-module,
(2) viewing $(\mathrm{Res}_H^G S)_T$ as a KG_T-module (see above), then

$$S \simeq \mathrm{Ind}_{G_T}^G (\mathrm{Res}_H^G S)_T .$$

The next statement is an immediate consequence of the above proposition.

Corollary 5.2.3 *Let S be an irreducible KG-module. Let H be a normal subgroup of G. Then*

(1) either there is a proper subgroup G' of G containing H, and an irreducible KG'-module S' such that $S = \mathrm{Ind}_{G'}^G S'$,
(2) or $\mathrm{Res}_H^G S$ is isotypic.

Application to nilpotent groups
The following proposition expresses the fact that a nilpotent group is an M-group.

Proposition 5.2.4 *Let G be a nilpotent finite group, and let K be a splitting field for G and all its subgroups. Let $S \in \mathrm{Irr}_K(H)$.*
   There exists a subgroup H of G and a 1-dimensional KH-module T such that $S = \mathrm{Ind}_H^G T$.

Proof We argue by induction on $|G|$. It is clear that the result is true if G is abelian, and so from now on we assume both that G is non abelian and that the conclusion of the proposition holds for any group of order strictly smaller than $|G|$.
   Let $\rho_S : G \to \mathrm{GL}(S)$ be the "structural morphism".

- Let G_0 be the kernel of ρ_S, so that ρ_S induces a natural structure of $K(G/G_0)$-module on S. If $G_0 \neq 1$, the induction hypothesis applied to the group G/G_0 insures the existence of a subgroup H/G_0 of G/G_0 and a degree one $K(H/G_0)$-module T such that $S = \mathrm{Ind}_{H/G_0}^{G/G_0} T$. With obvious notation, we thus have $S = \mathrm{Ind}_H^G T$.

- Now we assume $G_0 = 1$, *i.e.*, S is faithful for G. Let A be an abelian normal subgroup of G not contained in $Z(G)$ (see Appendix, Proposition A.2.4, (2)). Thus $\rho_S(A)$ is not contained in $\rho_S(Z(G))$, hence is not contained in $K\mathrm{Id}_S$. This shows that $\mathrm{Res}_A^G S$ is *not* isotypic. Now by Corollary 5.2.3, it follows that there exists a proper subgroup G' of G and an irreducible KG'-module S' such that $S = \mathrm{Ind}_{G'}^G S'$. We apply the induction hypothesis to G' to conclude.   □

The preceding proposition allows us to prove the following result of rationality, an anticipation of the Brauer–Schur theorem.

For G a finite group, we recall that e_G denotes the lcm of the orders of the elements of G, $\mu_{e_G}(K)$ denotes the group of e_G-th roots of unity in a splitting field of $X^{e_G} - 1$ over K.

Proposition 5.2.5 *Let G be a nilpotent finite group. If K is a characteristic zero field, any KG-module is rational over $\mathbb{Q}(\mu_{e_G}(K))$.*

Proof Notice first that if H is a subgroup of G,

$$\mathrm{Ind}_H^G \mathrm{Cha}_K(H) \subseteq \mathrm{Cha}_K(G) \text{ and } \mathrm{Res}_H^G \mathrm{Cha}_K(G) \subseteq \mathrm{Cha}_K(H).$$

Let L be a finite extension of $K(\mu_{e_G}(K))$ which is a splitting field for all subgroups of G. It suffices to prove that all LG-modules are rational over $\mathbb{Q}(\mu_{e_G}(K))$.

Notice that, whenever H is a subgroup of G, all degree one LH-modules are rational over $\mathbb{Q}(\mu_{e_G}(K))$.

Let $S \in \mathrm{Irr}_L(G)$. By Proposition 5.2.4, we know that there exists a subgroup H of G and a degree one LH-module T of H such that $S = \mathrm{Ind}_H^G(T)$. This shows that $\chi_S \in \mathrm{Ind}_H^G \mathrm{Cha}_{\mathbb{Q}(\mu_{e_G}(K))}(H)$, hence that $\chi_S \in \mathrm{Cha}_{\mathbb{Q}(\mu_{e_G}(K))}(G)$.   □

5.2.2 Induction and Restriction for Class Functions

Induction and restriction for class functions

We refer the reader to Notation 3.2.11 concerning class functions.

Definition 5.2.6 Let H be a subgroup of G. We define a linear map

$$\mathrm{Ind}_H^G : \begin{cases} \mathrm{CF}(H, \mathbb{Q}(\mu_{e_G}(K))) \to \mathrm{CF}(G, \mathbb{Q}(\mu_{e_G}(K))), \\ \beta \mapsto \left(\mathrm{Ind}_H^G \beta : s \mapsto \dfrac{1}{|H|} \displaystyle\sum_{\substack{g \in G \\ g^{-1}sg \in H}} \beta(g^{-1}sg)\right). \end{cases}$$

Remark 5.2.7 It is immediate to check that

$$\mathrm{Ind}_H^G \beta(s) = \sum_{\substack{r \in [G/H] \\ r^{-1}sr \in H}} \beta(r^{-1}sr).$$

Proposition 5.2.8 *Let Y be a KH-module, with character χ_Y. Then we have*

$$\chi_{\mathrm{Ind}_H^G Y} = \mathrm{Ind}_H^G \chi_Y.$$

Proof Since (see (5.1.5)) $\mathrm{Ind}_H^G Y = \bigoplus_{r \in [G/H]} r \otimes_{KH} Y$, we have

$$\chi_{\mathrm{Ind}_H^G Y}(g) = \bigoplus_{\substack{r \in [G/H] \\ grH=rH}} \mathrm{tr}_{(r \otimes_{KH} Y)/K}(g).$$

Since $grH = rH$ if and only if $r^{-1}gr \in H$, and since in that case

$$\mathrm{tr}_{(r \otimes_{KH} Y)/K}(g) = \mathrm{tr}_{Y/K}(r^{-1}gr),$$

the formula follows.   □

Proposition 5.2.9 (Frobenius Reciprocity) *Let H be a subgroup of G, $\alpha \in$ CF$(G, \mathbb{Q}(\mu_{e_G}(K)))$ and $\beta \in$ CF$(H, \mathbb{Q}(\mu_{e_G}(K)))$. Then*

$$\langle \alpha, \mathrm{Ind}_H^G \beta \rangle_G = \langle \mathrm{Res}_H^G \alpha, \beta \rangle_H.$$

Proof Indeed

$$\langle \alpha, \mathrm{Ind}_H^G \beta \rangle_G = \frac{1}{|G|} \sum_{g \in G} \alpha(g) \mathrm{Ind}_H^G \beta(g)^* = \frac{1}{|G|} \frac{1}{|H|} \sum_{\substack{g \in G \\ s \in G \\ s^{-1}gs \in H}} \alpha(g)\beta(s^{-1}gs)^*$$

$$= \frac{1}{|G|} \frac{1}{|H|} \sum_{\substack{s \in G \\ h \in H}} \alpha(shs^{-1})\beta(h)^* = \frac{1}{|G|} \frac{1}{|H|} \sum_{\substack{s \in G \\ h \in H}} \alpha(h)\beta(h)^*$$

$$= \frac{1}{|H|} \sum_{h \in H} \alpha(h)\beta(h)^* = \langle \mathrm{Res}_H^G \alpha, \beta \rangle_H.$$

                                                                                                □

Remark 5.2.10 We invite the reader to use the formula (see Proposition 5.1.15)

$$\mathrm{Hom}_{KG}(X, \mathrm{Ind}_H^G Y) \simeq \mathrm{Hom}_{KH}(\mathrm{Res}_H^G X, Y)$$

to give an alternative proof of Proposition 5.2.9.

The proof of the following proposition is easy and left to the reader.

Proposition 5.2.11 *Let H be a subgroup of G, $\alpha \in \mathrm{CF}(G, \mathbb{Q}(\mu_{e_G}(K)))$ and $\beta \in \mathrm{CF}(H, \mathbb{Q}(\mu_{e_G}(K)))$. Then*

$$\mathrm{Ind}_H^G((\mathrm{Res}_H^G \alpha)\beta) = \alpha \, \mathrm{Ind}_H^G \beta \,.$$

Remark 5.2.12 The reader may notice the connection of the preceding formula with Proposition 5.1.11.

Proposition 5.2.13 (Mackey Formula for Class Functions) *Let H and L be subgroups of G. Let $[L \backslash G / H]$ be a complete set of representatives of the double cosets of G modulo (L, H).*
    Whenever $\beta \in \mathrm{CF}(H, \mathbb{Q}_K^G)$,

$$\mathrm{Res}_L^G \mathrm{Ind}_H^G \beta = \sum_{r \in [L \backslash G / H]} \mathrm{Ind}_{L \cap {}^r H}^L \mathrm{Res}_{L \cap {}^r H}^{{}^r H} {}^r \beta \,.$$

Sketch of proof As for the Mackey formula for modules (see Theorem 5.1.19), this is an immediate consequence of Lemma 5.1.20.

Remark 5.2.14 The reader may also *deduce* Proposition 5.2.13 from Theorem 5.1.19.

We say that two KG-modules X and X' are *disjoint* if they have no isomorphic irreducible component in common, that is, if $\langle \chi_X, \chi_{X'} \rangle_G = 0$.

Proposition 5.2.15 (Irreducibility Criterion) *Let H be a subgroup of G and let T be a KH-module. The following assertions are equivalent:*

 (i) $\mathrm{Ind}_H^G T$ is an absolutely irreducible KG-module.
(ii) (a) T is an absolutely irreducible KH-module, and
        (b) whenever $g \in G \backslash H$, $\mathrm{Res}_{H \cap {}^g H}^H T$ and $\mathrm{Res}_{H \cap {}^g H}^{{}^g H} {}^g T$ are disjoint as $K(H \cap {}^g H)$-modules.

Proof This follows from Proposition 5.2.13 and from Proposition 5.2.9.   □

A result of Camille Jordan

Let us notice that combining Frobenius reciprocity Proposition 5.2.9 with the Mackey formula Proposition 5.2.13 provides the following proposition.

Proposition 5.2.16 *Let L and H be subgroups of G, let $\beta \in \mathrm{CF}(H, \mathbb{Q}_K^G)$ and $\gamma \in \mathrm{CF}(L, \mathbb{Q}_K^G)$. Then*

$$\langle \mathrm{Ind}_H^G \beta, \mathrm{Ind}_L^G \gamma \rangle_G = \sum_{r \in [L \backslash G / H]} \langle \mathrm{Res}_{L \cap {}^r H}^{{}^r H} {}^r \beta, \mathrm{Res}_{L \cap {}^r H}^L \gamma \rangle_{L \cap {}^r H} \,.$$

The next proposition is a particular case of the preceding proposition.

Proposition 5.2.17 *Let L and H be subgroups of G. Then*

$$\langle \chi_{K(G/H)}, \chi_{K(G/L)} \rangle_G = |L \backslash G / H|.$$

Remark 5.2.18 Applying the preceding proposition to the case $L = H$ provides another proof of Proposition 3.2.40.

The above proposition may be applied to prove a result of Camille Jordan.

Proposition 5.2.19 (Jordan) *Let G be a finite group and let H be a subgroup of G.*

(1) If H is a proper subgroup, there exists $g \in G$ which lies in no conjugate of H.
(2) If H is a proper subgroup,

$$\bigcup_{g \in G} gHg^{-1} \subsetneq G.$$

(3) If H meets every conjugacy class of G, then $H = G$.

Proof It is clear that the three assertions are equivalent. Let us prove (1).

Consider the linear representation $\mathbb{Q}(G/H)$ built from the transitive representation of G on the set G/H, and let χ be the corresponding character. Thus, for all $g \in G$, $\chi(g)$ is the number of fixed points of g acting on the set G/H, that is, the number of cosets $sH \in G/H$ such that $g \in sHs^{-1}$. Thus we must prove that there exists $g \in G$ such that $\chi(g) = 0$.

Since (see Lemma 5.1.4) $\chi = \mathrm{Ind}_H^G 1_H$, it follows from Frobenius reciprocity formula (see Proposition 5.2.9) that

$$\langle \chi, 1_G \rangle_G = \langle 1_H, 1_H \rangle_H = 1,$$

which implies

$$\frac{1}{|G|} \sum_{g \in G} \chi(g) = 1.$$

Since $\chi(g) \in \mathbb{N}$, $\chi(g) \neq 0$ implies $\chi(g) \geq 1$, and the above equality shows that there exists $g \in G$ such that $\chi(g) = 0$. $\qquad \square$

Remark 5.2.20 It can be shown, using the classification of finite simple groups, that in assertion (1) of the above Proposition 5.2.19, we may require g to be a p-element for some prime number p (see [FKS81]). Besides, using that result, it was proved by Malle–Navarro–Olsson [MNO00] that the element g in Theorem 3.7.13 may be chosen to be a p-element as well.

Gunter Malle (born 1960)

The next exercise proposes a "character–free" proof of Jordan's result.

Exercise 5.2.21 Let G be a finite group and let H be a proper subgroup of G. Prove that

$$\left| \bigcup_{g \in G} g H g^{-1} \right| \leq |G| - |G : H| + 1 .$$

HINT. Look for an upper bound for $\left| \bigcup_{g \in G} (g H g^{-1} \setminus \{1\}) \right|$.

5.2.3 Application of Induction to an Example

In this subsection, G is a finite group such that $G = A \rtimes H$, where A is abelian.

We shall compute the absolutely irreducible characters of such a group G.

Notice that this covers the case of the infinite family of reflection groups $G(de, e, r)$ (where $de \geq 2$), since

$$G(de, e, r) = D(de, e, r) \rtimes \mathfrak{S}_r ,$$

where $D(de, e, r)$ denotes the group of diagonal matrices with diagonal entries in $\boldsymbol{\mu}_{de}$ and determinant in $\boldsymbol{\mu}_d$.

Notation

We denote by K a field which is a splitting field for all subgroups of G.

We set $A^\vee := \operatorname{Hom}(A, K^\times)$. Thus $A^\vee$ is in natural bijection with $\operatorname{Irr}_K(A)$.

- For $\theta \in A^\vee$, we set

$$H_\theta := \{h \in H \mid {}^h\theta = \theta\} = \{h \in H \mid (\forall a \in A)(\theta(h^{-1}ah) = \theta(a))\} ,$$
$$G_\theta := A \rtimes H_\theta .$$

- The function $\tilde{\theta} : G_\theta \to K^\times$ defined by $\tilde{\theta}(ah) := \theta(a)$ (for all $a \in A$ and $h \in H$) is a group morphism.

Indeed,

$$\tilde{\theta}(a_1h_1a_2h_2) = \tilde{\theta}(a_1h_1a_2h_1^{-1}h_1h_2) = \theta(a_1h_1a_2h_1^{-1})$$
$$= \theta(a_1)\theta(h_1a_2h_1^{-1}) = \theta(a_1)\theta(a_2) = \theta(a_1a_2)$$
$$= \tilde{\theta}(a_1h_1)\tilde{\theta}(a_2h_2).$$

- For each $\rho \in \mathrm{Irr}_K(H_\theta)$, we denote by $\tilde{\rho}$ the irreducible character of G_θ defined by composition with the natural surjective morphism $G_\theta \twoheadrightarrow H_\theta = G_\theta/A$. Thus $\tilde{\theta} \cdot \tilde{\rho} \in \mathrm{Irr}_K(G_\theta)$.
- For each $\theta \in A^\vee$ and $\rho \in \mathrm{Irr}_K(H_\theta)$, we set

$$\chi_{(\theta,\rho)} := \mathrm{Ind}_{G_\theta}^G \tilde{\theta} \cdot \tilde{\rho}.$$

The group H acts on the set of pairs (θ, ρ). Indeed, for $h \in H$,

$${}^h(\theta, \rho) := ({}^h\theta, {}^h\rho) \in A^\vee \times \mathrm{Irr}_K(H_{{}^h\theta}).$$

Then for all $h \in H$, $\chi_{(\theta,\rho)} = \chi_{{}^h(\theta,\rho)}$.
Description of $\mathrm{Irr}_K(G)$

Proposition 5.2.22 *(1) For every pair* $(\theta, \rho) \in A^\vee \times \mathrm{Irr}_K(H_\theta)$, $\chi_{(\theta,\rho)} \in \mathrm{Irr}_K(G)$.
(2) $\chi_{(\theta,\rho)} = \chi_{(\theta',\rho')}$ *if and only if there exists* $h \in H$ *such that* $(\theta', \rho') = {}^h(\theta, \rho)$.
(3) $\mathrm{Irr}_K(G) = \{\chi_{(\theta,\rho)} \mid (\theta, \rho) \text{up to } H\text{-conjugation}\}$.

Proof 1. We prove

$$\mathrm{Ind}_A^G \theta = \sum_{\rho \in \mathrm{Irr}(H_\theta)} \rho(1)\chi_{(\theta,\rho)}. \tag{5.2.23}$$

Indeed, $\mathrm{Ind}_A^{G_\theta}\theta = \mathrm{Ind}_A^{G_\theta}((\mathrm{Res}_A^{G_\theta}\tilde{\theta})1_A)$. Hence by Proposition 5.2.11,

$$\mathrm{Ind}_A^{G_\theta}\theta = \tilde{\theta}\mathrm{Ind}_A^{G_\theta}(1_A) = \tilde{\theta}\widetilde{\rho_{H_\theta}^{\mathrm{reg}}} = \sum_{\rho \in \mathrm{Irr}(H_\theta)} \rho(1)\tilde{\theta}\tilde{\rho},$$

from which it follows that

$$\mathrm{Ind}_A^G \theta = \mathrm{Ind}_{G_\theta}^G \mathrm{Ind}_A^{G_\theta}\theta = \sum_{\rho \in \mathrm{Irr}(H_\theta)} \rho(1)\chi_{(\theta,\rho)}.$$

2. By Frobenius reciprocity, for all $\theta' \in \widehat{A}$,

$$\langle \mathrm{Res}_A^G \chi_{(\theta,\rho)}, \theta' \rangle_A = \langle \chi_{(\theta,\rho)}, \mathrm{Ind}_A^G \theta' \rangle_G = \langle \chi_{(\theta,\rho)}, \sum_{\rho' \in \mathrm{Irr}_K(H_{\theta'})} \rho'(1)\chi_{(\theta',\rho')} \rangle$$

$$= \sum_{\rho' \in \mathrm{Irr}_K(H_{\theta'})} \rho'(1)\langle \chi_{(\theta,\rho)}, \chi_{(\theta',\rho')} \rangle.$$

3. On the other hand,

$$\mathrm{Res}_A^G \chi_{(\theta,\rho)} = \mathrm{Res}_A^G \mathrm{Ind}_{G_\theta}^G \widetilde{\theta} \cdot \widetilde{\rho},$$

and it follows from Corollary 5.1.23 that

$$\mathrm{Res}_A^G \chi_{(\theta,\rho)} = \rho(1) \sum_{g \in [G/G_\theta]} {}^g\theta.$$

Applying 2 above, we get that, for all $\theta, \theta' \in A^\vee$ and all $\rho \in \mathrm{Irr}_K(H_\theta)$,

$$\sum_{\rho' \in \mathrm{Irr}_K(H_{\theta'})} \rho'(1) \langle \chi_{(\theta,\rho)}, \chi_{(\theta',\rho')} \rangle_G = \rho(1) \sum_{g \in [G/G_\theta]} \langle {}^g\theta, \theta' \rangle_A,$$

hence, since $G/G_\theta = H/H_\theta$,

$$\sum_{\rho' \in \mathrm{Irr}_K(H_{\theta'})} \rho'(1) \langle \chi_{(\theta,\rho)}, \chi_{(\theta',\rho')} \rangle_G = \begin{cases} \rho(1) & \text{if } \theta' = {}^h\theta \text{ for some } h \in H, \\ 0 & \text{if not.} \end{cases}$$

Since $\langle \chi_{(\theta,\rho)}, \chi_{(\theta',\rho')} \rangle_G \geq 0$ and $\langle \chi_{(\theta,\rho)}, \chi^h{}_{(\theta,\rho)} \rangle_G \geq 1$, the above equality shows that

$$\langle \chi_{(\theta,\rho)}, \chi_{(\theta',\rho')} \rangle_G = \begin{cases} 1 & \text{if } (\theta', \rho') = {}^h(\theta, \rho) \text{ for some } h \in H, \\ 0 & \text{if not,} \end{cases}$$

which establishes assertions (1) and (2).
    Since

$$\chi_G^{\mathrm{reg}} = \mathrm{Ind}_A^G \theta_A^{\mathrm{reg}} = \sum_{\theta \in A^\vee} \mathrm{Ind}_A^G \theta,$$

it follows from Eq. 5.2.23 in the above proof that

$$\chi_G^{\mathrm{reg}} = \sum_{\substack{\theta \in A^\vee \\ \rho \in \mathrm{Irr}_K(H_\theta)}} \rho(1) \chi_{(\theta,\rho)} = \sum_{\substack{\theta \in A^\vee/H \\ \rho \in \mathrm{Irr}_K(H_\theta)}} \chi_{(\theta,\rho)}(1) \chi_{(\theta,\rho)},$$

which shows (3).   □

More Exercises on Chap. 5

Exercise 5.2.24 Use Proposition 5.2.22 to recompute the character tables of $\mathfrak{S}_3$, $\mathfrak{A}_4$, $\mathfrak{S}_4$, D_8, D_{10}.

Exercise 5.2.25 Let p and ℓ be two prime numbers such that $p \equiv 1 \mod \ell$.

(1) Prove that, up to isomorphism, there are two groups of order $p\ell$.
(2) Compute the character tables of these two groups.

Exercise 5.2.26 Let q be a power of a prime p. Let $\mathbb{F}_q$ be a field with q elements. Let $G = \mathrm{SL}_2(\mathbb{F}_q)$ be the group of 2×2 matrices with entries in $\mathbb{F}_q$ and determinant 1. Let B be the *"Borel subgroup"* of G, defined by $B := \left\{ \begin{pmatrix} a & b \\ 0 & c \end{pmatrix} \mid a, b, c \in \mathbb{F}_q, \; ac = 1 \right\}$.

For $\xi : \mathbb{F}_q^\times \to \mathbb{C}^\times$ a group morphism, we get a group morphism

$$\theta_\xi : B \to \mathbb{C}^\times, \quad \begin{pmatrix} a & b \\ 0 & c \end{pmatrix} \mapsto \xi(a).$$

(1) Assume that $\xi^2 \neq 1$. Prove that $\mathrm{Ind}_B^G \theta_\xi$ is the character of an irreducible $\mathbb{C}G$-module.
(2) What is the value of $\mathrm{Ind}_B^G \theta_\xi$ on $\begin{pmatrix} a & b \\ d & c \end{pmatrix} \in \mathrm{SL}_2(\mathbb{F}_q)$?

Exercise 5.2.27 We recall that $G(de, e, r)$ $(d, e, r \geq 1)$ denotes the group of monomial matrices with nonzero entries in μ_{de} and product of nonzero entries in μ_d, so that

$$G(de, e, r) \simeq D(de, e, r) \rtimes \mathfrak{S}_r,$$

where $D(de, e, r)$ denotes the subgroup of diagonal matrices in $G(de, e, r)$.

(1) Use Proposition 5.2.22 to describe the character tables of the groups $G(4, 1, 3)$ and $G(4, 2, 3)$.
(2) Describe $\mathrm{Ind}_{G(4,2,3)}^{G(4,1,3)}(\chi)$ where χ runs over the set of irreducible characters of $G(4, 2, 3)$.

Exercise 5.2.28 Let $G = A \rtimes H$ where A is abelian. Prove directly (without using Proposition 5.2.22) that the number of conjugacy classes of G is equal to the number of orbits of H on the set of pairs (θ, ρ) where

- θ runs over the set of absolutely irreducible characters of A,
- ρ runs over the set of absolutely irreducible characters of H_θ.

Exercise 5.2.29 Let $G = H \times L$ be the direct product of two finite groups H and L. Let θ be a representation of H, and let ρ_L^{reg} be the regular representation of L. Prove that, for all $h \in H$ and $l \in L$,

$$\mathrm{Ind}_H^G(\theta)(h, l) = \theta(h)\rho_L^{\mathrm{reg}}(l).$$

Chapter 6
Brauer's Theorem and Some Applications

6.1 Artin's Theorem

As a preliminary for Brauer's characterization of generalized characters, we present a result of Artin, which characterizes (among all class functions) the elements of $\mathbb{Q}\mathrm{Cha}(G)$ (the $\mathbb{Q}$-vector space generated by $\mathrm{Cha}(G)$) in terms of their restrictions to cyclic groups.

We recall that, given a characteristic zero field K, we denote by $\mathrm{Cha}_K(G)$ the $\mathbb{Z}$-module with basis the family of characters of all irreducible KG-modules.

We denote by $\mathbb{Q}\mathrm{Cha}_K(G)$ the $\mathbb{Q}$-vector space of class functions generated by $\mathrm{Cha}_K(G)$.

Definition 6.1.1 Whenever A is a cyclic finite group, let us set

$$\gamma_A : A \to \mathbb{Z}, \ a \mapsto \begin{cases} |A| & \text{if } \langle a \rangle = A, \\ 0 & \text{if not.} \end{cases}$$

The following lemma is crucial for our proof of Artin's Theorem.

Lemma 6.1.2

(1) We have $\sum_{A \in \mathcal{CS}(G)} \mathrm{Ind}_A^G \gamma_A = |G| 1_G$,
(2) and $\gamma_A \in \mathrm{Cha}_\mathbb{Q}(A)$.

Proof
(1) By Definition 5.2.6, for $s \in G$,

$$\mathrm{Ind}_A^G \gamma_A(s) = \frac{1}{|A|} \sum_{\substack{g \in G \\ gsg^{-1} \in A}} \gamma_A(gsg^{-1})$$

© Springer Nature Singapore Pte Ltd. 2017
M. Broué, *On Characters of Finite Groups*, Mathematical Lectures from Peking University, https://doi.org/10.1007/978-981-10-6878-2_6

hence

$$\sum_{A \in \mathcal{CS}(G)} \mathrm{Ind}_A^G \gamma_A(s) = \sum_{A \in \mathcal{CS}(G)} \sum_{\substack{g \in G \\ \langle gsg^{-1}\rangle = A}} 1 = \sum_{g \in G} \sum_{\substack{A \in \mathcal{CS}(G) \\ \langle gsg^{-1}\rangle = A}} 1 = |G|.$$

(2) We prove the assertion by induction on $|A|$. It is true for $A = 1$. If $|A| > 1$, the preceding statement (1) may be re-written

$$\gamma_A + \sum_{\substack{B \in \mathcal{CS}(A) \\ B < A}} \mathrm{Ind}_B^A \gamma_B = |A| 1_A,$$

which shows that $\gamma_A \in \mathrm{Cha}(A)$.   □

Theorem 6.1.3 (Artin Theorem) *Let G be a finite group. Let $\mathcal{CS}(G)$ be the set of cyclic subgroups of G. Let K be a field.*

(1) Let $f : G \to K$ be a class function.

    (a) $|G| f = \sum_{A \in \mathcal{CS}(G)} \mathrm{Ind}_A^G (\gamma_A \mathrm{Res}_A^G f)$.

    (b) If, for all cyclic subgroup A of G, $\mathrm{Res}_A^G f \in \mathrm{Cha}_K(A)$, then $|G| f \in \mathrm{Cha}_K(G)$.

(2) $|G| \mathrm{Cha}_K(G) \subseteq \sum_{A \in \mathcal{CS}(G)} \mathrm{Ind}_A^G (\mathrm{Cha}_K(A)) \subseteq \mathrm{Cha}_K(G)$, and

$$\mathbb{Q}\mathrm{Cha}_K(G) = \sum_{A \in \mathcal{CS}(G)} \mathrm{Ind}_A^G (\mathbb{Q}\mathrm{Cha}_K(A)).$$

Proof
    (1)(a) results from Proposition 5.2.11 and from Lemma 6.1.2.
    (1)(b) is immediate by (a).
    (2) results also from (1)(a).   □

Remark 6.1.4 Let us remark that Lemma 4.2.9 has the following consequence for the group $\mathrm{Cha}_\mathbb{Q}(G)$ of generalized characters of $\mathbb{Q}G$-modules.

Proposition 6.1.5 *The $\mathbb{Q}$-vector space $\mathbb{Q}\mathrm{Cha}_\mathbb{Q}(G)$ generated by all characters of $\mathbb{Q}G$-modules is generated by the characters of permutation modules $\mathrm{Ind}_A^G 1_A = \mathbb{Q}(G/A)$, where A runs over the family of cyclic subgroups of G.*

Proof Indeed, it is enough to prove that if $f \in \mathbb{Q}\mathrm{Cha}_\mathbb{Q}(G)$ is orthogonal to all characters $\chi_{\mathbb{Q}(G/A)}$, then it is zero. But by Frobenius reciprocity,

$$\langle f, \mathrm{Ind}_A^G 1_A \rangle_G = \langle \mathrm{Res}_A^G f, 1_A \rangle_A,$$

and the result is then an immediate consequence of Lemma 4.2.9.   □

6.2 Brauer's Characterization of Characters

We recall (see Proposition 3.2.34) that, whenever K and L are two (commutative, characteristic zero) splitting fields for the finite group G, any choice of embeddings of K and L into a larger field provides an isomorphism between $\mathrm{Irr}_K(G)$ and $\mathrm{Irr}_L(G)$. In what follows, we shall denote by $\mathrm{Cha}(G)$ the group $\mathrm{Cha}_K(G)$ where K is a splitting field for G.

(!) This is a slight abuse of notation, since in general there is not a *unique* isomorphism between $\mathrm{Irr}_K(G)$ and $\mathrm{Irr}_L(G)$ (give an example!). But we shall only use this notation in a context where that abuse does not matter.

Definition 6.2.1 A (characteristic zero) field K is said *big enough* for G if it is a splitting field for all the subgroups of G.

One of the consequences of Brauer's Theorem proven below is that, for each finite group G, there is a smallest big enough field for G.

The presentation of Brauer's Theorem given here follows closely Chapter 10 of [Ser12].

Jean-Pierre Serre (born 1926)

6.2.1 Statement and First Consequences

Elementary groups.

Definition 6.2.2

(1) Let p be a prime. A finite group H is said to be *p-elementary* if $H = C \times P$ where C is a *cyclic* p'-group (i.e., of order prime to p) and P is a p-group.
(2) A group H is said to be *elementary* if there exists a prime p such that H is p-elementary.

Notice that any elementary subgroup is nilpotent, since it is the direct product of its various Sylow subgroups.

We denote by $\mathcal{ES}_p(G)$ (resp. $\mathcal{ES}(G)$) the set of all p-elementary subgroups of G (resp. the set of all elementary subgroups of G).

Statement.

For p a prime and n an integer, we denote by n_p the largest power of p which divides n, and we define $n_{p'}$ by the equality $n = n_p n_{p'}$.

Theorem 6.2.3 (Brauer's Theorem)

(1) Let p be a prime. Then

$$|G|_{p'}\mathrm{Cha}(G) \subset \sum_{H \in \mathcal{ES}_p(G)} \mathrm{Ind}_H^G \mathrm{Cha}(H) .$$

(2)

$$\mathrm{Cha}(G) = \sum_{H \in \mathcal{ES}(G)} \mathrm{Ind}_H^G \mathrm{Cha}(H) .$$

The proof will be given below.

First consequence: rationality of representations.

Theorem 6.2.4 (Brauer's Rationality Theorem) *Let G be a finite group, and let K be a splitting field for G. Then the field $\mathbb{Q}(\mu_{e_G}(K))$ is a big enough field for G.*

Proof Let $\chi \in \mathrm{Cha}(G)$. By Brauer's Theorem 6.2.3, (2), there exists a family (H, χ_H) where $H \in \mathcal{ES}(G)$ and $\chi_H \in \mathrm{Cha}(H)$, such that

$$\chi = \sum_{H \in \mathcal{ES}(G)} \mathrm{Ind}_H^G \chi_H .$$

Since H is nilpotent, Proposition 5.2.5 implies $\chi_H \in \mathrm{Cha}_{\mathbb{Q}(\mu_{e_G}(K))}(H)$, which implies that $\chi \in \mathrm{Cha}_{\mathbb{Q}(\mu_{e_G}(K))}(G)$, and we conclude by Proposition 4.2.5, (2). $\quad\square$

(!) **Attention** (!)

- There need not be a "smallest splitting field for G". More precisely, two subfields of a given field which are minimal splitting field for a finite group G need not be isomorphic: see Exercise 4.2.22.
- The field $\mathbb{Q}(\mu_{e_G})$ need not be a minimal splitting field for G.
  For example, one can prove that $\mathbb{Q}$ is a splitting field for the symmetric group $\mathfrak{S}_n$ for all $n > 0$.

Second consequence: characterization of characters.

Theorem 6.2.5 *Let K be a big enough field for G. For $f : G \to K$ a class function on G with values in K, the following assertions are equivalent.*

(i) $f \in \mathrm{Cha}_K(G)$.

(ii) *For each elementary subgroup H of G, $\mathrm{Res}_H^G f \in \mathrm{Cha}_K(H)$.*

Proof

(i)$\Rightarrow$(ii) is trivial. Let us prove (ii)$\Rightarrow$(i).

As above, by Brauer's Theorem there exists a family (ξ_H) where $H \in \mathcal{ES}(G)$ and $\xi_H \in \mathrm{Cha}(H)$, such that

$$1_G = \sum_{H \in \mathcal{ES}(G)} \mathrm{Ind}_H^G \xi_H \ .$$

By Proposition 5.2.11, it follows that

$$f = \sum_{H \in \mathcal{ES}(G)} \mathrm{Ind}_H^G (\xi_H \mathrm{Res}_H^G f) \,,$$

which makes obvious the statement to prove. $\qquad\square$

6.2.2 Proof of Brauer's Theorem

Principle of the proof.

- The second assertion of Brauer's Theorem 6.2.3 results from the first one.

    Indeed, let π_G denote the set of all primes which divide $|G|$. Since the family of integers $(|G|_{p'})_{p \in \pi_G}$ is relatively prime, there exists a family $(n_p)_{p \in \pi_G}$ of elements of $\mathbb{Z}$ such that $\sum_{p \in \pi_G} n_p |G|_{p'} = 1$.

    Let $\chi \in \mathrm{Cha}(G)$. By the first assertion of Brauer's Theorem, for each $p \in \pi_G$, $|G|_{p'}\chi \in \sum_{H \in \mathcal{ES}(G)} \mathrm{Ind}_H^G \mathrm{Cha}(H)$, thus

$$\chi = \sum_{p \in \pi_G} n_p |G|_{p'}\chi \in \sum_{H \in \mathcal{ES}(G)} \mathrm{Ind}_H^G \mathrm{Cha}(H) \,.$$

- In order to prove the first assertion of Brauer's Theorem 6.2.3, it suffices to prove

$$|G|_{p'} 1_G \in \sum_{H \in \mathcal{ES}_p(G)} \mathrm{Ind}_H^G \mathrm{Cha}(H) \,.$$

Indeed, if

$$|G|_{p'} 1_G = \sum_{H \in \mathcal{ES}_p(G)} \mathrm{Ind}_H^G \xi_H$$

for some family (ξ_H) where $\xi_H \in \mathrm{Cha}(H)$, then for all $\chi \in \mathrm{Cha}(G)$, it follows from Proposition 5.2.11 that

$$|G|_{p'}\chi = \sum_{H \in \mathcal{ES}_p(G)} \mathrm{Ind}_H^G(\xi_H \mathrm{Res}_H^G \chi).$$

- We define the following subring of the field $\mathbb{Q}_G$:

$$\mathbb{Z}_G := \mathbb{Z}[\boldsymbol{\mu}_{e_G}].$$

Remark 6.2.6 One may prove that $\mathbb{Z}_G$ is the ring of all algebraic integers of $\mathbb{Q}_G$.

We notice the following properties of $\mathbb{Z}_G$.

(1) Since $\mathbb{Z}_G$ is a torsionfree finitely generated $\mathbb{Z}$-module, it follows (see for example [Bro13], Proposition 2.179) that $\mathbb{Z}_G$ is a free $\mathbb{Z}$-module.
(2) Since $\mathbb{Z}_G$ consists of algebraic integers, $\mathbb{Q} \cap \mathbb{Z}_G = \mathbb{Z}$; this implies that $\mathbb{Z}_G/\mathbb{Z}$ is torsionfree, hence (*ibidem*) a free $\mathbb{Z}$-module. Whence the natural exact sequence

$$0 \to \mathbb{Z} \to \mathbb{Z}_G \to \mathbb{Z}_G/\mathbb{Z} \to 0$$

is split (see for example [Bro13], Proposition 2.38), hence there exists a submodule $\mathbb{Z}_G'$ of $\mathbb{Z}_G$ such that

$$\mathbb{Z}_G = \mathbb{Z} \oplus \mathbb{Z}_G',$$

and a surjective group morphism $\mathrm{pr}_G : \mathbb{Z}_G \twoheadrightarrow \mathbb{Z}$ inducing the identity on $\mathbb{Z}$. In particular, if p is a prime number, we have $p\mathbb{Z}_G \cap \mathbb{Z} = p\mathbb{Z}$.

- In order to prove the first assertion of Brauer's Theorem 6.2.3, it suffices to prove

$$|G|_{p'} 1_G \in \sum_{H \in \mathcal{ES}_p(G)} \mathbb{Z}_G \mathrm{Ind}_H^G \mathrm{Cha}(H). \qquad (6.2.7)$$

Indeed, assume equality 6.2.7. The morphism $\mathrm{pr}_G : \mathbb{Z}_G \to \mathbb{Z}$ extends naturally to a surjective morphism

$$\mathrm{pr}_G : \mathbb{Z}_G \mathrm{Cha}(G) \to \mathrm{Cha}(G)$$

which induces the identity on $\mathrm{Cha}(G)$. Applying pr_G to equality 6.2.7 gives then

$$|G|_{p'} 1_G \in \sum_{H \in \mathcal{ES}_p(G)} \mathrm{Ind}_H^G \mathrm{Cha}(H).$$

A key lemma.

Equality 6.2.7 will result from the following lemma.

Lemma 6.2.8 *(1) If $f \in \mathrm{CF}(G, \mathbb{Z})$ takes values all divisible by $|G|$, then*

$$f \in \sum_{A \in \mathcal{CS}(G)} \mathbb{Z}_G \mathrm{Ind}_A^G \mathrm{Cha}(A).$$

(2) Let p be a prime number. There exists

$$\psi \in \mathrm{CF}(G, \mathbb{Z}) \cap \sum_{H \in \mathcal{ES}_p(G)} \mathbb{Z}_G \mathrm{Ind}_H^G \mathrm{Cha}(H)$$

such that

$$\forall g \in G, \ \psi(g) \not\equiv 0 \mod p.$$

Indeed, let us prove that the above lemma implies Brauer's Theorem.

Let $\varphi : \mathbb{N} \setminus \{0\} \to \mathbb{N}$ denote the Euler function. For $m \in \mathbb{N}$ and $m \geq 2$, since the multiplicative group $(\mathbb{Z}/m\mathbb{Z})^\times$ has order $\varphi(m)$, for all n prime to m we have $n^{\varphi(m)} \equiv 1 \mod m$.

Let ψ be as in assertion (2) of Eq. (6.2.7). Then, by what precedes, for all $g \in G$,

$$\psi(g)^{\varphi(|G|_p)} \equiv 1 \mod |G|_p,$$

hence

$$|G|_{p'} \left(\psi(g)^{\varphi(|G|_p)} - 1 \right) \equiv 0 \mod |G|.$$

Now assertion (1) of Eq. (6.2.7) implies that

$$|G|_{p'} \left(\psi^{\varphi(|G|_p)} - 1 \right) \in \sum_{A \in \mathcal{CS}(G)} \mathbb{Z}_G \mathrm{Ind}_A^G \mathrm{Cha}(A),$$

hence in particular

$$|G|_{p'} \left(\psi^{\varphi(|G|_p)} - 1 \right) \in \sum_{H \in \mathcal{ES}_p(G)} \mathbb{Z}_G \mathrm{Ind}_H^G \mathrm{Cha}(H).$$

Since $\psi \in \sum_{H \in \mathcal{ES}_p(G)} \mathbb{Z}_G \mathrm{Ind}_H^G \mathrm{Cha}(H)$, we also have

$$|G|_{p'} \psi^{\varphi(|G|_p)} \in \sum_{H \in \mathcal{ES}_p(G)} \mathbb{Z}_G \mathrm{Ind}_H^G \mathrm{Cha}(H),$$

and so

$$|G|_{p'} 1_G \in \sum_{H \in \mathcal{ES}_p(G)} \mathbb{Z}_G \mathrm{Ind}_H^G \mathrm{Cha}(H),$$

which concludes the proof.

Proof of the key lemma.

- Let us first prove assertion (1) of Eq. (6.2.7).
  By Artin's Theorem 6.1.3, (1), we know that

$$|G| 1_G = \sum_{A \in \mathcal{CS}(G)} \mathrm{Ind}_A^G \gamma_A \, .$$

Thus if $f \in \mathrm{CF}(G, \mathbb{Z})$ is divisible by $|G|$, that is, if there exists $f_0 \in \mathrm{CF}(G, \mathbb{Z})$ such that $f = |G| f_0$, we have

$$f = |G| f_0 = \sum_{A \in \mathcal{CS}(G)} \mathrm{Ind}_A^G \left(\gamma_A \mathrm{Res}_A^G f_0 \right) \, .$$

Now, since $\gamma_A(a) = |A|$ if $\langle a \rangle = A$ and $\gamma_A(a) = 0$ if $\langle a \rangle \neq A$, we have

$$\gamma_A \mathrm{Res}_A^G f_0 = \sum_{\theta \in \mathrm{Irr}(A)} \langle \gamma_A \mathrm{Res}_A^G f_0, \theta \rangle_A \, \theta$$

$$= \sum_{\theta \in \mathrm{Irr}(A)} \sum_{\substack{a \in A \\ \langle a \rangle = A}} f_0(a) \theta(a^{-1}) \theta \, ,$$

which shows that $\gamma_A \mathrm{Res}_A^G f_0 \in \mathbb{Z}_G \mathrm{Cha}(A)$, and so that

$$f \in \sum_{A \in \mathcal{CS}(G)} \mathbb{Z}_G \mathrm{Ind}_A^G \mathrm{Cha}(A) \, .$$

- In order to prove assertion (2) of Eq. (6.2.7), we use the following technical lemma.

Lemma 6.2.9 *Let p be a prime number.*

(1) *If $f \in \mathrm{CF}(G, \mathbb{Z}) \cap \mathbb{Z}_G \mathrm{Cha}(G)$, then for all $g \in G$, whose p'-component is denoted by $g_{p'}$,*

$$f(g) \equiv f(g_{p'}) \mod p \, .$$

(2) *For any p'-element s of G, let us set $H_s := \langle s \rangle P$ for P a Sylow p-subgroup of the centralizer $C_G(s)$ of s. There exists*

$$\psi_s \in \mathrm{CF}(H_s, \mathbb{Z}) \cap \mathbb{Z}_G \mathrm{Cha}(H_s)$$

such that

   (a) $(\mathrm{Ind}_{H_s}^G \psi_s)(s) \not\equiv 0 \mod p$,
   (b) $(\mathrm{Ind}_{H_s}^G \psi_s)(t) \equiv 0 \mod p$ *whenever t is a p'-element of G not conjugate to s.*

Proof of Lemma 6.2.9 (1) We may assume $G = \langle g \rangle$ (and we do so). Thus the irreducible characters of G are group morphisms $G \to K^\times$.

Let $f = \sum_{\theta \in \mathrm{Irr}(G)} a_\theta \, \theta \in \mathrm{CF}(G, \mathbb{Z}) \cap \mathbb{Z}_G \mathrm{Cha}(G)$, so that, in particular, $a_\theta \in \mathbb{Z}_G$. We denote by p^∞ a large enough (i.e., larger than the contribution of p to the order of g) power of p.

Raising to the power p in any commutative ring R induces a ring endomorphism of R/pR, that is, $(x + y)^p \equiv x + y \mod pR$ for all $x, y \in R$ (see for example [Bro13], Exercise 1.28). Applying that remark to the case $R = \mathbb{Z}G$ yields

$$f(g)^{p^\infty} \equiv \sum_{\theta \in \mathrm{Irr}(G)} a_\theta^{p^\infty} \theta(g)^{p^\infty} \mod p\mathbb{Z}G. \qquad (*)$$

Since $g_{p'}^{p^\infty} = g^{p^\infty}$, for $\theta \in \mathrm{Irr}(G)$ we have

$$\theta(g)^{p^\infty} = \theta(g^{p^\infty}) = \theta(g_{p'}^{p^\infty}) = \theta(g_{p'})^{p^\infty},$$

and formula $(*)$ implies

$$f(g)^{p^\infty} \equiv f(g_{p'})^{p^\infty} \mod p\mathbb{Z}G. \qquad (**)$$

Since $f(g) \in \mathbb{Z}$ and $p\mathbb{Z}G \cap \mathbb{Z} = p\mathbb{Z}$, formula $(**)$ implies

$$f(g)^{p^\infty} \equiv f(g_{p'}^{p^\infty}) \mod p.$$

By Fermat's Theorem, we know that for $n \in \mathbb{Z}$, $n^p \equiv n \mod p$. This implies that

$$f(g) \equiv f(g_{p'}) \mod p.$$

(2) Let us set $A_s := \langle s \rangle$, and define

$$\psi_s := \sum_{\theta \in \mathrm{Irr}(A_s)} \theta(s^{-1})\theta.$$

By definition, $\psi_s \in \mathbb{Z}_G\mathrm{Cha}(A_s)$, and by the orthogonality relations,

$$\psi_s : \begin{cases} s \mapsto |A_s|, \\ t \mapsto 0 \text{ if } t \neq s. \end{cases}$$

By a slight abuse of notation, we still denote by ψ_s the function defined on $H_s = A_s P$ as the composition with the natural projection $H_s \twoheadrightarrow H_s/P = A_s$, that is

$$\psi_s : H_s \longrightarrow A_s \xrightarrow{\psi_s} K^\times.$$

Then

$$(\mathrm{Ind}_{H_s}^G \psi_s)(t) = \frac{1}{|H_s|} \sum_{\substack{g \in G \\ g^{-1}tg \in H_s}} \psi_s(g^{-1}tg),$$

hence

$$(\mathrm{Ind}_{H_s}^G \psi_s)(t) = 0 \text{ if } t \text{ is not conjugate to } s,$$

and

$$(\mathrm{Ind}_{H_s}^G \psi_s)(s) = \frac{1}{|H_s|} \sum_{\substack{g \in G \\ g^{-1}tg=s}} |A_s|$$

$$= \frac{1}{|H_s|}|C_G(s)||A_s| = \frac{|C_G(s)|}{|P|},$$

and since P is a Sylow p-subgroup of $C_G(s)$ this shows that

$$(\mathrm{Ind}_{H_s}^G \psi_s)(s) \not\equiv 0 \quad \mathrm{mod}\ p.$$

□

Now we can prove assertion (2) of the key Lemma 6.2.8, which will end the proof of Brauer's Theorem.

We denote by $G_{p'}$ the set of all p'-elements of G, and by $\mathrm{Cl}_G(G_{p'})$ a complete set of representatives of G-conjugacy classes of elements of $G_{p'}$. Let us set

$$\psi := \sum_{s \in \mathrm{Cl}_G(G_{p'})} \mathrm{Ind}_{H_s}^G \psi_s.$$

- By construction, $\psi \in \mathrm{CF}(G, \mathbb{Z}) \cap \sum_{H \in \mathcal{E}S_p(G)} \mathbb{Z}_G \mathrm{Cha}(H)$.
- By assertion (2) of Lemma 6.2.9, we see that, whenever $s \in G_{p'}$, we have $\psi(s) \not\equiv 0$ mod p.

Now it follows from assertion (1) of Lemma 6.2.9 that whenever $g \in G$, we have $\psi(g) \not\equiv 0$ mod p, which establishes assertion (2) of the key Lemma 6.2.8.

6.3 Fusion and Isometries

In this section, we apply induction and Brauer's characterization of characters in order to prove, in the following section, some results about the structure of some finite groups.

Throughout this section, K denotes a characteristic zero splitting field for the finite group G and all its subgroups.

6.3.1 π-elements and Class Functions

Let π be a set of prime numbers. We recall that for $g \in G$ we denote by g_π its π-component (see A.1.2), and by G_π the set of π-elements of G.

Notation 6.3.1 • We denote by $\mathrm{CF}_\pi(G, K)$ the subspace of K-valued class functions $f : G \to K$ such that, for all $g \in G$,

$$f(g) = f(g_\pi).$$

• If $u \in G_\pi$, we call π-*section* of u and we denote by $\mathrm{Sec}^G_\pi(u)$ the set of elements $g \in G$ such that g_π is G-conjugate to u.

Notice that in particular $\mathrm{Sec}^G_\pi(1) = G_{\pi'}$, where we denote by $G_{\pi'}$ the set of elements of G of order prime to the elements of π.

The proof of the following lemma is immediate.

Lemma 6.3.2

(1) The space $\mathrm{CF}_\pi(G, K)$ has for basis the family of characteristic functions of the π-sections of G.

(2) We have

$$\frac{|\mathrm{Sec}^G_\pi(u)|}{|G|} = \frac{|C_G(u)_{\pi'}|}{|C_G(u)|}.$$

Exercise 6.3.3 For $u \in G_\pi$, let $\mathrm{CF}(G, \mathrm{Sec}^G_\pi(u), K)$ denote the set of K-valued class functions on G which vanish outside of $\mathrm{Sec}^G_\pi(u)$.

(1) Prove that the map

$$C_G(u)_{\pi'} \to G \,, \quad s \mapsto us \,,$$

induces a bijection between the set of $C_G(u)$-conjugacy classes of $C_G(u)_{\pi'}$ and the set of G-conjugacy classes of $\mathrm{Sec}^G_\pi(u)$.

(2) Prove that the above map induces an isometry

$$\mathrm{CF}(C_G(u), C_G(u)_{\pi'}, K) \xrightarrow{\sim} \mathrm{CF}(G, \mathrm{Sec}^G_\pi(u), K) \,.$$

6.3.2 π-Control Subgroups

Let π be a set of primes.

Definition 6.3.4 We say that a subgroup H of a finite group G *controls the fusion of nilpotent π-subgroups* of G if

(C1) H contains at least one conjugate of each nilpotent π-subgroup of G,

(C2) if P is a nilpotent π-subgroup of G, and if $g \in G$ is such that $P \subset H$ and $gPg^{-1} \subset H$, then $g = hz$ for some $h \in H$ and $z \in C_G(P)$.

The next proposition provides an example of control of fusion.

William Burnside (1852–1927)

Proposition 6.3.5 (Burnside) *Let p be a prime number and G be a finite group whose Sylow p-subgroups are abelian. The normalizer $N_G(P)$ of a Sylow p-subgroup of G controls the fusion of p-subgroups of G.*

Proof Let us set $H := N_G(P)$. Condition (C1) is obviously (by the Sylow theorems) satisfied for H. Let us check condition (C2).

Assume that Q is a p-subgroup of G, and $g \in G$ is such that $Q \subset H$ and $gQg^{-1} \subset H$. Since P is the unique Sylow p-subgroup of H, hence the largest p-subgroup of H, it follows that both Q and gQg^{-1} are contained in P. Thus $Q \subset P \cap g^{-1}Pg$. Since P is abelian, both P and $g^{-1}Pg$ are Sylow p-subgroups of $C_G(Q)$. Hence by Sylow theorem applied to $C_G(Q)$, there exists $z \in C_G(Q)$ such that $zg^{-1}Pgz^{-1} = P$. Defining $h := gz^{-1}$, we see that $h \in H$ and $g = hz$.   □

If H controls the fusion of nilpotent π-subgroups of G, the inclusion $H \hookrightarrow G$ induces a bijection between the set of conjugacy classes of π-elements of H and the set of conjugacy classes of π-elements of G. Hence it induces a bijection between the set $\mathrm{Sec}_\pi(H)$ of π-sections of H and the set $\mathrm{Sec}_\pi(G)$ of π-sections of G, denoted by

$$\begin{cases} \mathrm{Sec}_\pi(H) \xrightarrow{\sim} \mathrm{Sec}_\pi(G) \ , \ T \mapsto T^G \ , \\ \mathrm{Sec}_\pi(G) \xrightarrow{\sim} \mathrm{Sec}_\pi(H) \ , \ S \mapsto S_H \ . \end{cases}$$

These bijections induce K-linear mutually inverse isomorphisms

$$\begin{cases} \mathrm{Pro}_H^G : \mathrm{CF}_\pi(H, K) \xrightarrow{\sim} \mathrm{CF}_\pi(G, K) \ , \ \gamma_T \mapsto \gamma_{T^G} \ , \\ \mathrm{Res}_H^G : \mathrm{CF}_\pi(G, K) \xrightarrow{\sim} \mathrm{CF}_\pi(H, K) \ , \ \gamma_S \mapsto \gamma_{S_H} \ . \end{cases}$$

Theorem 6.3.6 *Assume that H controls the fusion of nilpotent π-subgroups of G. The isomorphisms Pro_H^G and Res_H^G induce mutually inverse isometries*

$$\begin{cases} \mathrm{Pro}_H^G : \mathrm{CF}_\pi(H, K) \cap \mathrm{Cha}_K(H) \xrightarrow{\sim} \mathrm{CF}_\pi(G, K) \cap \mathrm{Cha}_K(G)\,, \\ \mathrm{Res}_H^G : \mathrm{CF}_\pi(G, K) \cap \mathrm{Cha}_K(G) \xrightarrow{\sim} \mathrm{CF}_\pi(H, K) \cap \mathrm{Cha}_K(H)\,. \end{cases}$$

Proof We shall first prove that Pro_H^G and Res_H^G are isometries between $\mathrm{CF}_\pi(H, K)$ and $\mathrm{CF}_\pi(G, K)$.

For this, it is enough to prove that, for each π-element $u \in H$,

$$\frac{|\mathrm{Sec}_\pi^G(u)|}{|G|} = \frac{|\mathrm{Sec}_\pi(u) \cap H|}{|H|}\,,$$

which, by Lemma 6.3.2, (2), is equivalent to proving

$$\frac{|C_G(u)_{\pi'}|}{|C_G(u)|} = \frac{|C_H(u)_{\pi'}|}{|C_H(u)|}\,.$$

This will result from the following lemma.

Lemma 6.3.7 *Let H be a subgroup of G which controls the fusion of nilpotent π-subgroups of G.*

(1) Whenever P is a nilpotent π-subgroups of H, then $C_H(P)$ controls the fusion of nilpotent π-subgroups of $C_G(P)$.
(2) We have

$$\frac{|G_{\pi'}|}{|G|} = \frac{|H_{\pi'}|}{|H|}\,.$$

Proof of Lemma 6.3.7 (1)(a) We prove that $C_H(P)$ contains a conjugate of every nilpotent π-subgroup of $C_G(P)$.

Let Q be a nilpotent π-subgroup of $C_G(P)$. Then PQ is a nilpotent π-subgroup of G, hence there exist $g \in G$ such that ${}^g(PQ) \subset H$. Since both P and ${}^gP$ are then contained in H, there exist $z \in C_G(P)$ and $h \in H$ such that $g = hz$. This implies ${}^zQ = {}^{h^{-1}g}Q \subset H$, hence ${}^zQ \subset C_H(P)$.

(1)(b) We prove that two nilpotent π-subgroups of $C_H(P)$ which are conjugate under $C_G(P)$ are already conjugate under $C_H(P)$.

Assume that Q is a nilpotent π-subgroup of $C_H(P)$, and that $x \in C_G(P)$ is such that ${}^xQ \subset C_H(P)$. Then both PQ and ${}^x(PQ)$ are nilpotent π-subgroups of H, hence there exist $z \in C_G(PQ) = C_{C_G(P)}(Q)$ and $y \in H$ such that $x = yz$. Since the last equality shows that $y \in C_H(P)$, this proves the desired property.

(2) We argue by induction on $|G|$.

Let R be a complete set of representatives for the H-conjugacy classes of π-elements of H ; by assumption, R is also a complete set of representatives for G-conjugacy classes of π-elements of G.

Set $R_0 := R \cap Z(G)$.

We have

$$1_G = \sum_{u \in R} \gamma_{\mathrm{Sec}^G_\pi(u)}$$

(where $\gamma_{\mathrm{Sec}^G_\pi(u)}$ denotes the characteristic function of $\mathrm{Sec}^G_\pi(u)$), hence

$$1 = \sum_{u \in R} \langle \gamma_{\mathrm{Sec}^G_\pi(u)}, \gamma_{\mathrm{Sec}^G_\pi(u)} \rangle_G = \sum_{u \in R} \frac{|C_G(u)_{\pi'}|}{|C_G(u)|} .$$

Similarly,

$$1 = \sum_{u \in R} \frac{|C_H(u)_{\pi'}|}{|C_H(u)|} .$$

The above equalities can be rewritten as follows:

$$|R_0| \frac{|G_{\pi'}|}{|G|} + \sum_{u \in R \setminus R_0} \frac{|C_G(u)_{\pi'}|}{|C_G(u)|} = |R_0| \frac{|H_{\pi'}|}{|H|} + \sum_{u \in R \setminus R_0} \frac{|C_H(u)_{\pi'}|}{|C_H(u)|} .$$

Since, for all $u \in R \setminus R_0$, $|C_G(u)| < |G|$, the induction hypothesis, assertion (1), and the above formula, prove assertion (2). $\qquad \square$

Now we prove that the maps Res^G_H and Pro^G_H induce mutually inverse isomorphisms between $\mathrm{CF}_\pi(H, K) \cap \mathrm{Cha}_K(H)$ and $\mathrm{CF}_\pi(G, K) \cap \mathrm{Cha}_K(G)$.

It is clear that Res^G_H sends $\mathrm{CF}_\pi(G, K) \cap \mathrm{Cha}_K(G)$ into $\mathrm{CF}_\pi(H, K) \cap \mathrm{Cha}_K(H)$.

Let us check that Pro^G_H sends $\mathrm{CF}_\pi(H, K) \cap \mathrm{Cha}_K(H)$ into $\mathrm{CF}_\pi(G, K) \cap \mathrm{Cha}_K(G)$.

Let $\chi \in \mathrm{CF}_\pi(H, K) \cap \mathrm{Cha}_K(H)$. Since $\mathrm{Pro}^G_H(\chi) \in \mathrm{CF}_\pi(G, K)$, it suffices to prove that $\mathrm{Pro}^G_H(\chi) \in \mathrm{Cha}_K(G)$. By Brauer's Theorem 6.2.3, it suffices to prove that, whenever E is a nilpotent subgroup of G, then $\mathrm{Res}^G_E \mathrm{Pro}^G_H(\chi) \in \mathrm{Cha}_K(E)$.

Since E is nilpotent, $E = E_\pi \times E_{\pi'}$, where E_π (resp. $E_{\pi'}$) is a π-group (resp. π'-group). Since H controls the fusion of nilpotent π-subgroups of G, we may assume (and we do so) that $E_\pi \subset H$. Then $\mathrm{Res}^G_E \mathrm{Pro}^G_H(\chi) = (\mathrm{Res}^G_{E_\pi} \chi) 1_{E_{\pi'}}$, which obviously belongs to $\mathrm{Cha}_K(E)$. $\qquad \square$

6.4 Some Fundamental Theorems about Finite Groups

As an application of Theorem 6.3.6, we present some fundamental theorems about finite groups.

6.4.1   Existence of a Normal π-Complement

The reader may prove as an exercise that, given π a set of prime numbers, the subgroup of G generated by all normal π-subgroups of G is still a normal π-subgroup of G. Thus this subgroup is *the largest normal π-subgroup of G*. It is usually denoted by $O_\pi(G)$.

The next theorem was proved by Frobenius in 1907 for the case where π is a singleton.

Theorem 6.4.1 *Assume that π is a set of primes and that H is a π-subgroup of G which controls the fusion of nilpotent π-subgroups of G. Then H "has a normal π-complement", that is (denoting by π' the complement of π relative to the set of prime numbers)*

$$G = H O_{\pi'}(G).$$

Before proving Theorem 6.4.1, let us give two consequences concerning "p-local theory" of finite groups.

Corollary 6.4.2 (Burnside) *Let P be a Sylow p-subgroup of G. If $N_G(P) = C_G(P)$ (in other words, if P is contained in the center of $N_G(P)$), then $G = P O_{p'}(G)$.*

Proof of Corollary 6.4.2 It follows from the hypothesis that P is abelian. By Proposition 6.3.5, we then see that $N_G(P)$, hence $C_G(P)$, controls the fusion of p-subgroups. But then there is "no fusion", and so P controls the fusion of p-subgroups. Thus the assertion follows from Theorem 6.4.1.   □

Corollary 6.4.3 *Assume that G has a cyclic Sylow 2-group P. Then $G = P O_{2'}(G)$.*

Proof By Corollary 6.4.2, it suffices to prove that $N_G(P) = C_G(P)$. Since $N_G(P)/C_G(P)$ is isomorphic to a subgroup of $\mathrm{Aut}(P)$, and since $P \subset C_G(P)$, this results from the fact that the automorphism group of a cyclic 2-group is a 2-group, since $\varphi(2^n) = 2^{n-1}$.

Remark 6.4.4 The above corollary shows in particular that a nonabelian simple group cannot have a cyclic Sylow 2-subgroup.

Actually, by the Feit–Thompson theorem which proves that any group of odd order is solvable, it follows from Corollary 6.4.3 that any finite group with a cyclic Sylow 2-subgroup is solvable.

Proof of Theorem 6.4.1 Let K be a characteristic zero field which is splitting for all subgroups of G. Since H is a π-group, we have

$$\mathrm{CF}_\pi(H, K) \cap \mathrm{Cha}_K(H) = \mathrm{Cha}_K(H),$$

hence by Theorem 6.3.6 we have an isometry

$$\mathrm{Pro}_H^G : \mathrm{Cha}_K(H) \xrightarrow{\sim} \mathrm{CF}_\pi(G, K) \cap \mathrm{Cha}_K(G).$$

If T is an absolutely irreducible KH-module, it follows that $\mathrm{Pro}_H^G(\chi_T)$ is an element of $\mathrm{Cha}_K(G)$ such that

$$\langle \chi_T, \chi_T \rangle = 1 \ \text{ and } \ \mathrm{Pro}_H^G(\chi_T)(1) = \chi_T(1) > 0,$$

hence $\mathrm{Pro}_H^G(\chi_T)$ is the character of an absolutely irreducible KG-module. Moreover, by definition of Pro_H^G and by Proposition 3.2.2, we see that every π'-element of G belongs to the kernel of that irreducible character $\mathrm{Pro}_H^G(\chi_T)$.

Let us now prove, denoting by $G_{\pi'}$ the set of π'-elements of G, that

$$\bigcap_{T \in \mathrm{Irr}_K(H)} \ker\left(\mathrm{Pro}_H^G(\chi_T)\right) = G_{\pi'} . \tag{ker}$$

Indeed, let $g \in G \setminus G_{\pi'}$. We shall prove that there is $T \in \mathrm{Irr}_K(H)$ such that $g \notin \ker\left(\mathrm{Pro}_H^G(\chi_T)\right)$. We may assume (and we do so) that $h := g_\pi \in H$. Since $h \neq 1$, there exists $T \in \mathrm{Irr}_K(H)$ such that $h \notin \ker(\chi_T)$, i.e., $\chi_T(h) \neq \chi_T(1)$. It follows that $\mathrm{Pro}_H^G(\chi_T)(g) \neq \mathrm{Pro}_H^G(\chi_T)(1)$, that is, $g \notin \ker\left(\mathrm{Pro}_H^G(\chi_T)\right)$.

The equality (ker) shows that $G_{\pi'} = O_{\pi'}(G)$. Since then $O_{\pi'}(G)$ contains all the Sylow subgroups for all primes in π', we see that $|O_{\pi'}(G)| = |G|_{\pi'}$. Similarly, since H contains at least one Sylow p-subgroup for all $p \in \pi$, we see that $|H| = |G|_\pi$. This implies that the order of the subgroup $HO_{\pi'}(G)$ of G is equal to the order of G, hence that $G = HO_{\pi'}(G)$. □

6.4.2 π-trivial Intersection Subgroups

Definition 6.4.5 Let π be a set of prime numbers, and let H be a subgroup of G. We say that H is a π-*trivial intersection subgroup of* G if

(1) whenever $p \in \pi$, p divides $|H|$,
(2) for all $s \in G \setminus N_G(H)$,

$$H \cap sHs^{-1} \text{ is a } \pi' - \text{group.}$$

Remark 6.4.6 If H is a π-trivial intersection subgroup of G, then whenever P is a nontrivial π-subgroup of H, we have $N_G(P) \subset N_G(H)$.

Indeed, whenever $g \in N_G(P)$, we have $g^{-1}Pg \subset H$, hence $P \subset H \cap gHg^{-1}$.

The next proposition is easily proved in the case where π is a singleton. But we shall need it in the general case.

Proposition 6.4.7 *Let H be a π-trivial intersection subgroup of G.*

(1) The following assertions are equivalent.

    (i) H contains a conjugate of every π-subgroup of G.

(ii) $N_G(H)/H$ is a π'-group.

(iii) Whenever P is a nontrivial π-subgroup of H, then $C_H(P)$ is a normal subgroup of $C_G(P)$ and the group $C_G(P)/C_H(P)$ is a π'-group.

(2) If the above assertions are satisfied, then

(a) $N_G(N_G(H)) = N_G(H)$,

(b) $N_G(H)$ is a π-trivial intersection subgroup of G,

(c) $N_G(H)$ controls the fusion of nilpotent π-subgroups of G.

Proof

(i)$\Rightarrow$(ii) : Since H contains a Sylow p-subgroup of G for each $p \in \pi$, $|G : H|_\pi = 1$, and so $N_G(H)/H$ is a π'-group.

(ii)$\Rightarrow$(iii) : Since $C_H(P) = C_G(P) \cap H$, and since (by Remark 6.4.6) $C_G(P)$ normalizes H, we see that $C_H(P)$ is a normal subgroup of $C_G(P)$. Moreover, $C_G(P)/C_H(P) = C_G(P)/H \cap C_G(P) = HC_G(P)/H$ is a subgroup of $N_G(H)/H$, hence a π'-group by (ii).

(iii)$\Rightarrow$(i) : We first prove that, whenever $p \in \pi$, H contains a Sylow p-subgroup of G.

Indeed, let $p \in \pi$ and let P be a Sylow p-subgroup of G such that $P \cap H$ is a Sylow p-subgroup of H. Since $Z(P) \subset C_G(P \cap H)$ and since $P \cap H \neq 1$, we have $C_G(P \cap H)/C_H(P \cap H)$ is a π'-group and so $Z(P) \subset H$. Since $P \subset C_G(Z(P))$ and since $C_G(Z(P))/C_H(Z(P))$ is a π'-group (note that $Z(P) \neq 1$), it follows that $P \subset H$.

Now let E be a nontrivial nilpotent π-subgroup of G. Let p be a prime which divides $|E|$. With obvious notation, we have $E = E_p \times E_{p'}$ and $E_p \neq 1$. By what precedes, there is $g \in G$ such that $gE_pg^{-1} \subset H$. Since $gE_{p'}g^{-1} \subset C_G(gE_pg^{-1})$, it follows from the hypothesis that $gE_{p'}g^{-1} \subset H$, hence $gEg^{-1} \subset H$.

Now we prove (2).

By (ii), every π-subgroup of $N_G(H)$ is contained in H, hence for all $s \in G$, every π-subgroup of $N_G(H) \cap sN_G(H)s^{-1}$ is contained in $H \cap sHs^{-1}$. If $s \notin N_G(H)$, $H \cap sHs^{-1}$ is a π'-group, hence we see that for all $s \in G \setminus N_G(H)$, $N_G(H) \cap sN_G(H)s^{-1}$ is a π'-group.

This proves (a) and (b). Moreover, if P is a nontrivial nilpotent π-subgroup of $N_G(H)$ and $g \in G$ such that $gPg^{-1} \in N_G(H)$, we have $P \subset N_G(H) \cap g^{-1}N_G(H)g$, which shows that $g \in N_G(H)$ and proves (c). $\qquad\square$

6.4.3   Frobenius Groups

The main theorem.

Theorem 6.4.8 *Let G be a finite group.*

(1) The following assertions are equivalent.

(i) *There exists a proper nontrivial subgroup H of G such that*

$$\forall g \in G \setminus H , \ H \cap gHg^{-1} = 1 .$$

(ii) *There exists a nontrivial proper normal subgroup N of G such that*

$$\forall n \in N, n \neq 1 , \ C_G(n) \subset N .$$

(2) *If the above assertions are satisfied, then*

  (a) *N is nilpotent,*
  (b) *N is the only subgroup of G which satisfies (ii),*
  (c) *the only subgroups of G which satisfy (i) are the subgroups conjugate to H,*
  (d) *G = NH and N ∩ H = 1, thus $G \simeq N \rtimes H$,*
  (e) $N = \{1\} \cup \left(G - \bigcup_{g \in G} gHg^{-1} \right)$,
  (f) $|N|$ *and* $|H|$ *are relatively prime.*

Remark 6.4.9 We shall not give a self-contained proof of the above theorem. Indeed,
• the proof of (ii)⇒(i) uses the Schur–Zassenhaus Theorem, which is only stated in the Appendix (see A.4.1),
• the proof of (2)(a) and (2)(b) uses a deep theorem of J.G. Thompson, which we cannot prove here.

Definition 6.4.10 A group which satisfies properties (1) of the above theorem is called a *Frobenius group*.

• The unique normal subgroup N defined by (ii) is then called the *Frobenius kernel* of the Frobenius group G.
• The subgroups conjugate to H defined by (i) are called the *Frobenius complements* of the Frobenius group G.

Proof of Theorem 6.4.8 (i)⇒(ii), (2)(d), (2)(e) : Let π_H be the set of primes which divide $|H|$. Since $N_G(H) = H$ and H is a π_H-trivial intersection subgroup of G, it follows from item (2)(c) of Proposition 6.4.7 that H controls the fusion of nilpotent π_H-subgroups of G. Then Theorem 6.4.1 implies that there is a normal π_H-complement N, i.e., $G \simeq N \rtimes H$, which proves (2)(d) (provided (ii) holds, which is proved below). By hypothesis, we then have

$$\left| \bigcup_{n \in N} nHn^{-1} \right| = |N|(|H| - 1) + 1 ,$$

which implies

$$G = N \cup \bigcup_{n \in N} nHn^{-1} ,$$

and proves (2)(e) (provided (ii) holds, which is proved below).

Let $n \in N, n \neq 1$. If $C_G(n)$ were not contained in N, it would contain ghg^{-1} for some $h \in H, h \neq 1$ and $g \in G$. Replacing n by $g^{-1}ng$ we see that we may (and we do) assume $g = 1$. Since H is a π_H-trivial intersection subgroup, it follows from Remark 6.4.6 that $C_G(h)$ must be contained in H, hence cannot contain n. This proves (ii).

(ii)$\Rightarrow$(i), (2)(f) : Let π_N be the set of primes which divide $|N|$. We apply here item (1) of Proposition 6.4.7.

Since N is normal in G, N is clearly a π_N-trivial intersection subgroup. Assertion (1)(iii) of Proposition 6.4.7 is satisfied since by hypothesis $C_G(M) = C_N(M)$ whenever M is a nontrivial subgroup of N. Thus by (1)(i) of the same proposition, $N_G(N)/N = G/N$ is a π'_N-group, proving (2)(f).

By item (1) of the Schur–Zassenhaus Theorem (see Theorem A.4.1), there exists a subgroup H of G such that $G \simeq N \rtimes H$. Let $n \in N$. We first prove that $C_H(n) = H \cap nHn^{-1}$.

It is clear that $C_H(n) \subset H \cap nHn^{-1}$. Conversely, if $h \in H \cap nHn^{-1}$, then $[n^{-1}, h] = n^{-1}hnh^{-1} \in H \cap N$ hence $[n^{-1}, h] = 1$ and $h \in C_H(n)$.

It follows from(ii) that, for each $n \in N, n \neq 1$, we have $H \cap nHn^{-1} = 1$, which proves (i).

Only (2)(a), (b) and (c) remain to be proved.

First we notice that by (1)(ii), a nontrivial element of H acts on N by conjugation without fixing any nontrivial element. Thompson's Theorem (for a proof, see [Hup67], Kap. V, Haupsatz 8.14) implies that N is nilpotent, i.e., (2)(a).

The uniqueness of N, i.e., (2)(b), follows from the next lemma.

Lemma 6.4.11 *Assume there are subgroups H and N of G satisfying the properties (i) and (ii) of Theorem 6.4.8, (1).*

(1) If N' is any normal subgroup of G, then either $N' \subset N$ or $N \subset N'$.
(2) If moreover N' also satisfies property (ii), then $N' = N$.

Proof (1) Let $N' \lhd G$. Assume $N' \not\subset N$. We shall prove that $N \subset N'$. Since $G = N \cup \bigcup_{n \in N} nHn^{-1}$, there is $n \in N$ such that $N' \cap nHn^{-1} \neq 1$, which implies $N' \cap H \neq 1$. Let p be a prime number dividing $|N' \cap H|$, and let P be a Sylow p-subgroup of $N' \cap H$. Since $P \subset H$, it follows from Remark 6.4.6 that $N_G(P) \subset H$ and so $N_{N'}(P) \subset N' \cap H$. Thus P is a Sylow p-subgroup of $N_{N'}(P)$, which implies (by the local characterisation of Sylow subgroups, see A.3.1) that P is a Sylow p-subgroup of N'. Now by the Frattini argument (see A.3.2) we see that $G = N'N_G(P)$, hence $G = N'H$. In particular, G/N' is isomorphic to a quotient of H, hence has order prime to N, which shows that the image of N in G/N' is trivial, that is $N \subset N'$.

(2) Now assume moreover that for all $n \in N'$, $C_G(n) \subset N'$. By what has been proved above, since N and N' play analogous roles, we may as well assume (which we do) that $N' \subset N$. Since N is nilpotent by Thompson's Theorem, we have $Z(N) \cap N' \neq 1$, which implies by hypothesis $C_G(Z(N) \cap N') \subset N'$, and finally

$$N \subset C_G(Z(N) \cap N') \subset N'.$$

$\square$

Now we prove (2)(c). Assume that H' is a subgroup of G satisfying (1)(i), i.e., for all $g \notin H'$, $H' \cap gH'g^{-1} = 1$. Then, by what we have proven above we know that there is a normal $\pi_{H'}$-complement N, which satisfies (1)(i) and is unique by what precedes. By the second item of the Schur–Zassenhaus Theorem (see A.4.1), since N is nilpotent, we see that H' is conjugate to H. $\qquad\square$

Irreducible characters of a Frobenius group.

To describe the set of irreducible characters of a Frobenius group, we need the following lemma.

Lemma 6.4.12 *Let G be a Frobenius group, with kernel N and complement H.*

(1) The complement H acts semi-regularly on the set $\mathrm{Cl}(N) - \{1\}$ of nontrivial conjugacy classes of N, that is, if $h \in H$ stabilizes a nontrivial conjugacy class of N, then $h = 1$.

(2) $|H|$ divides $|\mathrm{Cl}(N)| - 1$ and $|N| - 1$.

Proof

(1) Let $h \in H$, and let $n \in N$, $n \neq 1$ whose conjugacy class in N is fixed under h. Thus $hnh^{-1} = mnm^{-1}$ for some $m \in N$. Then $h^{-1}m \in C_G(n)$, hence $h^{-1}m \in N$, which implies $h \in N$ and so $h = 1$.

(2) Since $\mathrm{Cl}(N) - \{1\}$ is the disjoint union of orbits of H, and since by (1) each orbit has cardinality $|H|$, we see that $|H|$ divides $|\mathrm{Cl}(N)| - 1$. Moreover, with obvious notation,

$$|N| - 1 = \sum_{C \in \mathrm{Cl}(N)-\{1\}} |C| = \sum_{C \in (\mathrm{Cl}(N)-\{1\})/H} |H||C|,$$

which shows that $|H|$ divides $|N| - 1$. $\qquad\square$

Theorem 6.4.13 *Let G be a Frobenius group with complement H and kernel N.*

(1) The family of absolutely irreducible characters of G consists of

   (a) the family $\left(\chi_\xi := \mathrm{Pro}_H^G \xi\right)_\xi$ where ξ runs over the set of absolutely irreducible characters of H,

   (b) the family $(\chi_\theta := \mathrm{Ind}_N^G \theta)_\theta$ where θ runs over a complete set of representatives $[\mathrm{Irr}(N)/H]$ of the H-orbits on the set of nontrivial absolutely irreducible characters of N.

(2) Here are the numbers of various absolutely irreducible characters:

$$\begin{cases} \left|\{\chi_\xi\}_{\xi \in \mathrm{Irr}(H)}\right| = |\mathrm{Irr}(H)| \\[2mm] \left|\{\chi_\theta\}_{\theta \in [\mathrm{Irr}(N)/H], \, \theta \neq 1_N}\right| = \dfrac{1}{|H|}\big(|\mathrm{Irr}(N)| - 1\big) \end{cases}$$

(3) *Let π_N denote the set of prime numbers which divide $|N|$, let $\mathrm{Ind}_N^G \mathrm{Cha}(N)^0$ denote the subgroup of the group $\mathrm{Ind}_N^G \mathrm{Cha}(N)$ consisting of generalized characters χ such that $\chi(1) = 0$. Then*

$$\mathrm{CF}_{\pi_N}(G) \cap \mathrm{Cha}(G) = \mathbb{Z}1_G + \mathrm{Ind}_N^G \mathrm{Cha}(N)^0 \, .$$

Proof Given the structure of G, an irreducible character χ of G is equal to $\mathrm{Pro}_H^G(\xi)$ for some irreducible character ξ of H if and only if its kernel contains N. Hence it suffices to prove that, if χ is an irreducible character of G such that $N \not\subseteq \ker \chi$, then χ is induced from a nontrivial character of N.

To prove this, by Proposition 5.2.2, it suffices to check that, whenever θ is a nontrivial irreducible character of N, the fixator of θ in G is equal to N. In other words, we have to check that no nontrivial element of H fixes a nontrivial irreducible character of N. Let $h \in H, h \neq 1$. The number of irreducible characters of N fixed by h equals the trace of h acting on the space $\mathrm{CF}(N, K)$ of class functions on N, hence equals 1 by item (1) of Lemma 6.4.12.

Thus we have proved that H acts semi-regularly on $\mathrm{Irr}(N) - \{1_N\}$. In particular the number of orbits of H on $\mathrm{Irr}(N) - \{1_N\}$ is $(|\mathrm{Irr}(N)| - 1)/|H|$, which proves (1) and (2).

By item (2)(e) of Theorem 6.4.8, we know that $G = N \cup G_{\pi_N'}$, which implies that $\mathrm{CF}_{\pi_N}(G, K) \cap \mathrm{Cha}(G)$ is the set of all generalized characters which are constant on $G_{\pi_N'}$, that is, the set of all generalized characters whose restriction to H is a multiple of the trivial character 1_H of H.

Let us set $\mathrm{Irr}(N)^{\sharp} := \mathrm{Irr}(N) - \{1_N\}$, and let

$$\chi = \sum_{\xi \in \mathrm{Irr}(H)} a_\xi \chi_\xi + \sum_{\theta \in [\mathrm{Irr}(N)^{\sharp}/H]} b_\theta \chi_\theta \qquad (\chi)$$

be a generalized character of G. We have

$$\mathrm{Res}_H^G \chi = \sum_{\xi \in \mathrm{Irr}(H)} a_\xi \xi + \Big(\sum_{\theta \in [\mathrm{Irr}(N)^{\sharp}/H]} b_\theta \theta(1) \Big) \xi_H^{\mathrm{reg}}$$

$$= \sum_{\xi \in \mathrm{Irr}(H)} \Big(a_\xi + \Big(\sum_{\theta \in [\mathrm{Irr}(N)^{\sharp}/H]} b_\theta \theta(1) \Big) \xi(1) \Big) \xi \, .$$

Thus $\chi \in \mathrm{CF}_{\pi_N}(G, K)$ if and only if, for all nontrivial $\xi \in \mathrm{Irr}(H)$, we have

$$a_\xi + \Big(\sum_{\theta \in [\mathrm{Irr}(N)^{\sharp}/H]} b_\theta \theta(1) \Big) \xi(1) = 0 \, . \qquad (*)$$

Let γ_N be the characteristic function of N, that is

$$|H|\gamma_N = \sum_{\xi \in \mathrm{Irr}(H)} \xi(1)\chi_\xi = \mathrm{Ind}_N^G 1_N \,.$$

Using equation (*) in the equality (χ) above, we get

$$\chi = -\sum_\theta b_\theta \theta(1)|H|\gamma_N + \sum_\theta b_\theta \chi_\theta = \sum_\theta b_\theta \big(\chi_\theta - \chi_\theta(1)\gamma_N\big)\,.$$

This proves (3). $\square$

More Exercises on Chap. 6

In what follows, G is a finite group, p is a prime number, and we assume that $|G| = p^a m$ where m is prime to p.

Let S be an irreducible $\mathbb{C}G$-module.

Exercise 6.4.14

(1) Define the class function $\chi_S^{(p)}$ on G by $\chi_S^{(p)}(g) := \chi_S(g_{p'})$. Prove that $\chi_S^{(p)} \in$ Cha(G).

(2) Define the class function $\widetilde{\chi}_S^{(p)}$ on G by

$$\widetilde{\chi}_S^{(p)}(g) := \begin{cases} \chi_S(g) & \text{if } g_p = 1, \\ 0 & \text{if } g_p \neq 1. \end{cases}$$

Prove that $p^a \widetilde{\chi}_S^{(p)} \in$ Cha(G).

Exercise 6.4.15 ["p-DEFECT ZERO" CHARACTERS.] We use the notation $\widetilde{\chi}_S^{(p)}$ defined in the preceding exercise.

Let H be a nilpotent subgroup of G, so $H = P \times Q$ where P is a p-group and Q is p'-group.

(1) Prove that for each irreducible character ζ of H,

$$|P| \langle \text{Res}_H^G \widetilde{\chi}_S^{(p)}, \zeta \rangle_H \in \mathbb{Z}.$$

(2) Let $\omega_S : Z\mathbb{C}G \to \mathbb{C}$ be the algebra morphism associated with S. For all $s \in Q$, we denote by C_s its conjugacy class in G. We recall that $\omega_S(SC_s) = \chi_S(s)|C_s|/\chi_S(1)$ is an algebraic integer.

(a) Prove that

$$|H| \langle \text{Res}_H^G \widetilde{\chi}_S^{(p)}, \zeta \rangle_H = |Q| \langle \text{Res}_Q^G \chi_S, \text{Res}_Q^H \zeta \rangle_Q,$$

and deduce that $|P| \langle \text{Res}_H^G \widetilde{\chi}_S^{(p)}, \zeta \rangle_H \in \mathbb{Z}$.

(b) Prove that

$$|H| \langle \text{Res}_H^G \widetilde{\chi}_S^{(p)}, \zeta \rangle_H = \frac{\chi_S(1)}{|G|} \sum_{s \in Q} \omega_S(SC_s) \zeta(x)^* |C_G(s)|,$$

then, since $P \subset C_G(s)$, prove that

$$(|G||Q|/\chi_S(1)) \langle \text{Res}_H^G \widetilde{\chi}_S^{(p)}, \zeta \rangle_H = \sum_{s \in Q} \omega_S(SC_s) \zeta(x)^* |C_G(s) : P|.$$

(c) Notice (cf. (a)) that $\langle \mathrm{Res}_H^G \widetilde{\chi}_S^{(p)}, \zeta \rangle_H \in \mathbb{Q}$ to deduce from what precedes that
$(|G||Q|/\chi_S(1)) \langle \mathrm{Res}_H^G \widetilde{\chi}_S^{(p)}, \zeta \rangle_H \in \mathbb{Z}$.

(3) We shall now prove the following result.

Proposition 6.4.16 ["p-defect zero" characters.] *Assume that $p^a \mid \chi_S(1)$. Then whenever $g \in G$ is such that $g_p \neq 1$, we have $\chi_S(g) = 0$.*

(a) We assume $p^a \mid \chi_S(1)$. Deduce from what precedes that $\langle \mathrm{Res}_H^G \widetilde{\chi}_S^{(p)}, \zeta \rangle_H \in \mathbb{Z}$, hence $\mathrm{Res}_H^G \widetilde{\chi}_S^{(p)} \in \mathrm{Cha}(H)$.
(b) Prove that
$$\widetilde{\chi}_S^{(p)} \in \mathrm{Cha}(G).$$

(c) Prove Proposition 6.4.16.

Exercise 6.4.17

(1) Let N be a finite group with an involutive automorphism τ which has no fixed point but 1 on N.

  (a) Prove that for all $n \in N$, $\tau(n) = n^{-1}$.
       HINT. Consider the map $N \to N$, $n \mapsto n^{-1}\tau(n)$ and prove that is is injective, hence bijective.
  (b) Prove that N is abelian.

(2) Let G be a Frobenius group with kernel N, complement H of even order. Prove that N is abelian.
     NOTE. This is an easy and particular case of Thompson's Theorem since an abelian group is nilpotent.

Chapter 7
Graded Representation and Characters

7.1 Graded Vector Spaces, Algebras, Modules

Here we limit ourselves, for simplicity and since we do not need more, to some finiteness conditions, like imposing each homogeneous component to be finite dimensional, and free modules to have countable rank.

7.1.1 Graded Vector Spaces

First definitions.
Let k be a field.

Definition 7.1.1 (1) We call *graded k-vector space* any k-vector space of the form

$$V = \bigoplus_{n=-\infty}^{n=\infty} V_n$$

where

- for each n, V_n is a *finite dimensional k-vector space,*
- $V_n = 0$ whenever $n < N$ for some integer N (i.e., "for n small enough"), and we write $V = \bigoplus_{n>-\infty}^{\infty} V_n$.

(2) A graded vector space morphism $V \to V'$ is a linear map $f : V \to V'$ such that, for each $n \in \mathbb{Z}$, we have $f(V_n) \subset V'_n$.

For each integer n, the non-zero elements of V_n are called *homogeneous of degree* n. The degree of a homogeneous element x is denoted by $\deg(x)$.
If $x = \sum_n x_n$ where $x_n \in V_n$, then the element x_n is called the *homogeneous component of degree n* of x.

© Springer Nature Singapore Pte Ltd. 2017
M. Broué, *On Characters of Finite Groups*, Mathematical Lectures from Peking University, https://doi.org/10.1007/978-981-10-6878-2_7

A *graded subspace* V' of V is a subspace such that

$$V' = \bigoplus_{n=-\infty}^{\infty} (V' \cap V_n) \,,$$

thus it is

- a subspace V' such that the homogeneous components of its elements belong to V',
- a graded space and a subspace of V such that the natural injection $V' \hookrightarrow V$ is a morphism of graded vector spaces.

Example 7.1.2 Let X be an indeterminate. Then $k[X]$ is a graded k-vector space, $k[X^2]$ is a graded subspace.

Graded dimension.
    Let q be an indeterminate.
    We set $\mathbb{Z}((q)) := \mathbb{Z}[[q]][q^{-1}]$, which is the ring of formal Laurent series with coefficients in $\mathbb{Z}$.
    Let V be a graded vector space. *The graded dimension* of V is the element of $\mathbb{Z}((q))$ defined by

$$\operatorname{grdim}_k(V) := \sum_{n=-\infty}^{\infty} [V_n : k]q^n \,.$$

Example 7.1.3 Let X be an indeterminate. Then

$$\operatorname{grdim}_k(k[X]) = \frac{1}{1-q} \quad \text{and} \quad \operatorname{grdim}_k(k[X^2]) = \frac{1}{1-q^2} \,.$$

Elementary constructions.
- *Direct sum* – If V and V' are two graded vector spaces, the graded vector space $V \oplus V'$ is defined by the condition $(V \oplus V')_n := V_n \oplus V'_n$.
    We have

$$\operatorname{grdim}_k(V \oplus V') = \operatorname{grdim}_k(V) + \operatorname{grdim}_k(V') \,.$$

- *Tensor product* – If V and V' are two graded vector spaces, the graded vector space $V \otimes V'$ is defined by the condition $(V \otimes V')_n := \bigoplus_{i+j=n} V_i \otimes V'_j$.
    Note that this is well defined since $V_n = 0$ for n "small enough".
    We have

$$\operatorname{grdim}_k(V \otimes V') = \operatorname{grdim}_k(V)\operatorname{grdim}_k(V') \,.$$

- *Shift* – If V is a graded vector space and m is an integer, the graded vector space $V[m]$ is defined by the condition $V[m]_n := V_{m+n}$.

We have

$$\operatorname{grdim}_k(V[m]) = q^{-m} \operatorname{grdim}_k(V).$$

Example 7.1.4 • If $Y_1, \ldots, Y_r$ are algebraically independent homogeneous elements of $k[X_1, \ldots, X_r]$ of degree $d_1, d_2, \ldots, d_r$ respectively, then we have $k[Y_1, \ldots, Y_r] \cong k[Y_1] \otimes_k \cdots \otimes_k k[Y_r]$ and

$$\operatorname{grdim}_k(k[Y_1, \ldots, Y_r]) = \frac{1}{(1 - q^{d_1})(1 - q^{d_2}) \cdots (1 - q^{d_r})}.$$

• If V has dimension 1 and is generated by an element of degree d, then we have $V \cong k[-d]$, and $\operatorname{grdim}_k(V) = q^d$.

• For V a finite dimensional graded vector space, V has a basis consisting of homogeneous elements. If $m := [V : k]$ and if $(e_1, \ldots, e_m)$ is the family of degrees of the elements of such a basis, we have

$$\operatorname{grdim}_k(V) = q^{e_1} + \cdots + q^{e_m}.$$

7.1.2 Graded Algebras and Modules

First definitions.

Definition 7.1.5 (1) A graded k-algebra A is a k-algebra which, as a k-vector space, is graded — thus $A = \bigoplus_{n > -\infty}^{\infty} A_n$, — and such that

• $A_n = 0$ if $n < 0$, and $A_0 = k$,
• $A_m A_n \subset A_{m+n}$.

(2) A graded A-module M is

• an A-module and a graded k-vector space — thus

$$M = \bigoplus_{n > -\infty}^{n=\infty} M_n,$$

• such that for all integers m and n, $A_m M_n \subset M_{m+n}$.

Notice that a graded algebra is naturally a graded module over itself.
We leave it to the reader to check the following obvious lemma.

Lemma 7.1.6 (1) *If A is a graded k-algebra (resp. a graded k-algebra which is finitely generated as a k-algebra), there exists a family (resp. finite family) of homogeneous elements $(a_i)_{i \in I}$ of A, all of degree at least 1, such that $A = k[(a_i)_{i \in I}]$.*

(2) *If M is a graded A-module (resp. a graded A-module which is finitely generated as an A-module) there exists a family (resp. finite family) of homogeneous
elements $(x_j)_{j \in J}$ of M such that $M = \sum_{j \in J} Ax_j$.*

Definition 7.1.7 (1) A *graded A-module morphism* is an A-module morphism
which is a morphism of graded k-vector spaces.
(2) A *graded submodule N* of a graded A-module M is an A-submodule such that
the natural injection is a morphism of graded k-vector spaces, i.e., such that
$N = \bigoplus_n (N \cap M_n)$.

Lemma 7.1.8 *Let N be an A-submodule of a graded A-module M. The following
assertions are equivalent.*

(i) *N is a graded submodule of M.*
(ii) *Each homogeneous component of an element of N belongs to N.*
(iii) *There exists a family $(y_j)_{j \in J}$ of homogeneous elements of N such that $N = \sum_{j \in J} Ay_j$.*

Proof (i)$\Rightarrow$(ii)$\Rightarrow$(iii) are straightforward and left to the reader. Let us prove (iii)$\Rightarrow$(i).
It is enough to prove that

$$N = \bigoplus_n (N \cap M_n) \text{ and even that } N = \sum_n (N \cap M_n).$$

Let us denote by e_j the degree of y_j. For any $a \in A$, we have $a = \sum a_m$ for some
$a_m \in A_m$, hence, given $j \in J$, $ay_j = \sum_m a_m y_j \in \sum_m N \cap M_{m+e_j} = \sum_n N \cap M_n$.
For any $y \in N$, there is a finite family $(a_j)_{j \in J}$ with $a_j \in A$ such that $y = \sum_{j \in J} a_j y_j$, and by what precedes it follows that $y \in \sum_n N \cap M_n$. $\quad\square$

Lemma 7.1.9 *Let N be a graded A-submodule of a graded A-module M. Set
$(M/N)_n := (M_n + N)/N$. Then*

$$M/N = \bigoplus_n (M/N)_n$$

is a graded A-module.

Proof It is clear that $M/N = \sum_n (M/N)_n$, and also that

$$A_m (M/N)_n \subset (M/N)_{m+n}.$$

To prove that $\sum_n (M/N)_n = \bigoplus_n (M/N)_n$, it is enough to prove that, given $x = \sum_n x_n \in M$ such that $x \in N$, then for all n we have $x_n \in N_n$, which is immediate
by the definition of graded submodule. $\quad\square$

The proof of the following proposition is left to the reader.

Proposition 7.1.10 *Let $f : M \to M'$ be a morphism of graded A-modules.*

(1) *The image of f is a graded A-submodule of M'.*
(2) *The kernel of f is a graded A-submodule of M.*
(3) *f induces an isomorphism of graded A-modules*

$$M/\ker(f) \xrightarrow{\sim} \mathrm{im}\,(f).$$

Graded ideals.

A graded (or "homogeneous") ideal of A is a graded submodule of A, seen as graded module over itself.

Thus the following properties are easy consequences of Lemmas 7.1.8 and 7.1.9.

Proposition 7.1.11 *Let A be a graded k-algebra.*

(1) *Let $\mathfrak{a}$ be a proper ideal of A. The following conditions are equivalent:*

   (i) $\mathfrak{a}$ *is a graded proper ideal,*
   (ii) $\mathfrak{a} = \bigoplus_{n \geq 1}(\mathfrak{a} \cap A_n),$
   (iii) *for all $a \in \mathfrak{a}$, each homogeneous component of a belongs to $\mathfrak{a}$,*
   (iv) $\mathfrak{a}$ *is generated (over A) by a family of homogeneous elements each of degree at least 1.*

(2) *Let $\mathfrak{a}$ be a graded ideal of A, and let M be a graded A-module. We denote by $\mathfrak{a}M$ the submodule of M generated by the elements am for $a \in \mathfrak{A}$ and $m \in M$. Then*

   (a) $\mathfrak{a}M$ *is a graded submodule of M, where $(\mathfrak{a}M)_n$ is the k-vector space generated by all elements am where $\deg(a) + \deg(m) = n$,*
   (b) *$A/\mathfrak{a}$ is a graded algebra, and $M/\mathfrak{a}M$ is a graded $A/\mathfrak{a}$-module.*

It is immediate that

$$\mathfrak{M} := \bigoplus_{n=1}^{\infty} A_n$$

is then the largest graded ideal of A, hence called *the maximal graded ideal of A*.

The following properties are easy to check.

Lemma 7.1.12

(M1) *The injection $k \hookrightarrow A$ induces an isomorphism $k \xrightarrow{\sim} A/\mathfrak{M}$.*
(M2) $A = k[\mathfrak{M}].$
(M3) *For a family $(a_i)_{i \in I}$ of homogeneous elements of A each of degree at least 1, the following assertions are equivalent:*

   (i) $A = k[(a_i)_{i \in I}],$
   (ii) $\mathfrak{M} = \sum_{i \in I} A a_i.$

Tensor, symmetric and exterior algebras.

Throughout this paragraph, V denotes a k-vector space of finite dimension r.

The proof of the following proposition is left to the reader, who may find helps or hints in many places in the literature.

Proposition 7.1.13 *Denote by $(e_1, \ldots, e_r)$ a basis of V.*

(1) *For all $n \in \mathbb{N}$, let us denote by $T(V)_n$ the subspace of the tensor algebra $T(V)$ generated by all the elementary n-tensors $v_1 \otimes_k \cdots \otimes_k v_n$ for $v_i \in V$. Then*

    (a) *the family $(e_{i_1} \otimes_k \cdots \otimes_k e_{i_n})_{1 \leq i_1, \ldots, i_n \leq r}$ is a basis of $T(V)_n$,*

    (b) *$T(V) = \bigoplus_n T(V)_n$ is a graded k-algebra, whose graded dimension is*

$$\mathrm{grdim}_k(T(V)) = \frac{1}{1 - rq}.$$

(2) *The structure of graded k-algebra of $T(V)$ induces on $S(V)$ a structure of graded k-algebra. Moreover,*

    (a) *the family $(e_1^{d_1} \cdots e_r^{d_r})_{\sum_i d_i = n}$ is a basis of $S(V)_n$,*

    (b) *the graded dimension of the graded k-algebra $S(V)$ is*

$$\mathrm{grdim}_k(S(V)) = \frac{1}{(1 - q)^r}.$$

(3) *The structure of graded k-algebra of $T(V)$ induces on $\Lambda(V)$ a structure of graded k-algebra. Moreover,*

    (a) *the family $(e_{i_1} \wedge \cdots \wedge e_{i_n})_{1 \leq i_1 < \cdots < i_n \leq r}$ is a basis of $\Lambda(V)_n$,*

    (b) *the graded dimension of the graded k-algebra $\Lambda(V)$*

$$\mathrm{grdim}(\Lambda(V)) = (1 + q)^r.$$

7.1.3 Nakayama's Lemma

Theorem 7.1.14 (Graded Nakayama's Lemma) *Let A be a graded k-algebra, with maximal graded ideal $\mathfrak{M}$.*

Let M be a graded A-module.

(1) *If $\mathfrak{M}M = M$, then $M = 0$.*

(2) *If M' is a graded submodule of M, then $M' = M$ if and only if $M = M' + \mathfrak{M}M$.*

(3) *If $f : M \to N$ is a morphism of graded A-modules which induces a surjection from M onto $N/\mathfrak{M}N$, then f is surjective.*

(4) *A system $(x_j)_{j \in J}$ of homogeneous elements of M is a generating system for M if and only if its image in $M/\mathfrak{M}M$ is a generating system of the k-vector space $M/\mathfrak{M}M$.*

Proof (1) Assume $M \neq 0$. Let m be a nonzero element of M with minimal degree. Then $m \notin \mathfrak{M}M$.

(2) By (1),

$$M' = M \iff \mathfrak{M}(M/M') = M/M'.$$

Since $\mathfrak{M}(M/M') = (\mathfrak{M}M + M')/M'$,

$$\mathfrak{M}(M/M') = M/M' \iff \mathfrak{M}M + M' = M.$$

(3) Apply (2) replacing M by N and M' by $f(M)$.
(4) Apply (2) where M' is the submodule generated by $(x_j)_{j \in J}$. $\qquad\square$

Corollary 7.1.15 *If M is a graded A-module, all its minimal generating systems as an A-module have the same cardinality, which is the dimension of $M/\mathfrak{M}M$ over k.*

Corollary 7.1.16 *Let $(u_i)_{i \in I}$ be a family of homogeneous elements of A with positive degrees. The following assertions are equivalent:*

(i) $A = k[(u_i)_{i \in I}]$,
(ii) $\mathfrak{M} = \sum_{i \in I} A u_i$,
(iii) $\mathfrak{M}/\mathfrak{M}^2 = \sum_{i \in I} k u_i$.

7.1.4 Free Graded Modules

Throughout this paragraph, A denotes a graded k-algebra, with maximal graded ideal $\mathfrak{M}$.

Definition 7.1.17 A graded A-module M is said to be *free graded* if there exist a finite set I, a family $(e_i)_{i \in I}$ of elements of $\mathbb{Z}$, and an isomorphism of graded A-modules

$$\bigoplus_{i \in I} A[-e_i] \xrightarrow{\sim} M.$$

Remark 7.1.18 A graded k-vector space is free graded.
  Indeed, let $i \in I$ be a family of integers satisfying the condition

$$\mathrm{grdim}_k V = \sum_{i \in I} q^{e_i}.$$

Then $V \cong \bigoplus_{i \in I} k[-e_i]$.

  The proof of the following lemma is immediate and left to the reader.

Lemma 7.1.19 *Let M be a free graded A-module, and let* $(e_i)_{i \in I}$ *be such that there is an isomorphism* $\bigoplus_{i \in I} A[-e_i] \xrightarrow{\sim} M$ *. Let us denote by* x_i *the image of the unit element of A (in degree* e_i *) under the above isomorphism, and let us denote by* $\bar{x}_i$ *its image in* $M/\mathfrak{M}M$.

(1) *The A-linear map*

$$A \otimes_k M/\mathfrak{M}M \to M \ , \ a \otimes_k \bar{x}_i \mapsto a x_i$$

  is an isomorphism of graded A-modules.
(2) $\mathrm{grdim}_k(M) = \mathrm{grdim}_k(A) \left(\sum_{i \in I} q^{e_i} \right)$ *.*
(3) *The family* $(e_i)_{i \in I}$ *is uniquely determined by M (up to permutations).*

The next result shows that a summand of a free graded module is also free graded.

If I is a finite set and A is a ring, we denote by A^I the free A-module with basis indexed by I.

Theorem 7.1.20 *Let A be a graded k-algebra, and let M be a graded A-module.*

Assume that, as a (not necessarily graded) A-module, M is a summand of a free A-module, i.e., there exist a finite set I, an A-module N, and an isomorphism of A-module $A^I \xrightarrow{\sim} M \oplus N$.

Then M is free graded.

Remark 7.1.21 The above statement remains true "without graduation" under more restrictive hypothesis – this is Serre's conjecture, proved by Quillen and Suslin (see for example [Lan65]): let $A := k[X_1, \ldots, X_r]$ and let M be a *finitely generated* A-module which is a summand of a free A-module. Then M is free.

Proof of Theorem 7.1.20 Since $M/\mathfrak{M}M$ is a graded k-vector space, it is a free graded k-vector space (see Remark 7.1.18): we may assume given a family of integers $(e_i)_{i \in I}$ and an isomorphism

$$\bar{\sigma} : \bigoplus_{i \in I} k[-e_i] \xrightarrow{\sim} M/\mathfrak{M}M .$$

Thus there exists a family $(x_i)_{i \in I}$ of homogeneous elements of M (with $\deg x_i = e_i$) whose image in $M/\mathfrak{M}M$ is a basis, which implies (by Theorem 7.1.14) that $(x_i)_{i \in I}$ generates M, and this provides a surjective morphism of graded A-modules

$$\sigma : \bigoplus_{i \in I} A[-e_i] \twoheadrightarrow M .$$

We shall prove that the above morphism is an isomorphism.

Let us denote by M' the kernel of the above morphism. We shall prove that $M' = 0$.

By hypothesis, there exist a set I, an A-module N, and two A-linear maps

$$A^I \xrightarrow{\ \pi \ } M \qquad \text{and} \qquad M \xrightarrow{\ \iota \ } A^I$$

such that $\pi \cdot \iota = \mathrm{Id}_M$.

The map which sends the i-th element of the canonical basis of A^I onto a reciprocal image by σ of its image under π induces a morphism of A-modules

$$\tilde{\pi} : A^I \to \bigoplus_{i \in I} A[-e_i]$$

such that the following diagram commutes:

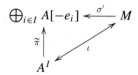

Define $\sigma' := \tilde{\pi} \cdot \iota$, so that the following diagram commutes:

Then

$$\sigma \cdot \sigma' = \sigma \cdot \tilde{\pi} \cdot \iota = \pi \cdot \iota = \mathrm{Id}_M \,,$$

which shows that σ is split (as a morphism of A-modules). Thus the following short exact sequence

$$0 \longrightarrow M' \longrightarrow \bigoplus_{i \in I} A[-e_i] \overset{\sigma}{\longrightarrow} M \longrightarrow 0$$

is split (as a sequence of A-modules). This implies that the horizontal sequences of the following commutative diagram are exact (and split) — while the vertical arrows are surjective:

Since $\bar{\sigma}$ is an isomorphism, $M'/\mathfrak{M}M' = 0$. By Nakayama's Theorem (7.1.14, (1)), this implies $M' = 0$. Hence σ is an isomorphism. $\qquad\square$

7.1.5 *Polynomial Algebras and Noether Parameters*

Degrees and subalgebras.

Let $S = k[v_1, v_2, \ldots, v_r]$ be a polynomial graded algebra over the field k, where $(v_1, v_2, \ldots, v_r)$ is a family of algebraically independent, homogeneous elements, with degrees respectively $e_1, e_2, \ldots, e_r$. Assume $e_1 \le e_2 \le \cdots \le e_r$.

Let $(u_1, u_2, \ldots, u_r)$ be a family of homogeneous elements with degrees d_1, $d_2, \ldots, d_r$ such that $d_1 \le d_2 \le \cdots \le d_r$.

Lemma 7.1.22 *Assume that $(u_1, u_2, \ldots, u_r)$ is algebraically independent.*

(1) *For all i $(1 \le i \le r)$, we have $e_i \le d_i$.*
(2) *We have $e_i = d_i$ for all i $(1 \le i \le r)$ if and only if $S = k[u_1, \ldots, u_r]$.*

Proof of Lemma 7.1.22 (1) Let i be such that $1 \le i \le r$. The family $(u_1, \ldots, u_i)$ is algebraically independent, hence it cannot be contained in $k[v_1, \ldots, v_{i-1}]$. Hence there exist $j \ge i$ and $l \le i$ such that v_j does appear in u_l. It follows that $e_j \le d_l$, hence $e_i \le e_j \le d_l \le d_i$.
(2) We know that $\operatorname{grdim} S = (\prod_{i=1}^{r}(1 - q^{e_i}))^{-1}$. Thus it suffices to prove that $\prod_{i=1}^{r}(1 - q^{e_i}) = \prod_{i=1}^{r}(1 - q^{d_i})$ if and only if $e_i = d_i$ for all i $(1 \le i \le r)$, which is left as an exercise.   □

Definition 7.1.23 By Lemma 7.1.22, we see in particular that the family $(e_1, e_2, \ldots, e_r)$ (with $e_1 \le e_2 \le \cdots \le e_r$) is uniquely determined by S. Such a family is called *the family of characteristic degrees* of S.

Transcendence degree and Krull dimension.

Let A be a commutative finitely generated graded k-algebra. For simplicity, we assume that A is an integral domain. We denote by K its field of fractions.

We recall (see for example [Lan65], Chap. VIII) that the *transcendence degree* of K over k is the integer d defined by the following condition:

There exist d elements $x_1, \ldots, x_d$ of K such that

(Tr1) The family $(x_1, \ldots, x_d)$ is algebraically independent over k (thus $k(x_1, \ldots, x_d)$ is *purely transcendental* over k),
(Tr2) K is an algebraic extension of $k(x_1, \ldots, x_d)$.

We call *Krull dimension* of A, and we denote by $\operatorname{Krdim}(A)$, the transcendence degree of K.

Remark 7.1.24 The Krull dimension of A may be defined, without assuming that A is an integral domain, as the largest integer d such that there exists a sequence $\mathfrak{P}_1 \subsetneq \cdots \subsetneq \mathfrak{P}_{d-1}$ of prime ideals in A.

Noether parameters.

Definition 7.1.25 A *Noether system of parameters* of a finitely generated graded k-algebra A is a family $(x_1, \ldots, x_r)$ of homogeneous elements in A such that

(N1) $(x_1, \ldots, x_r)$ is algebraically independent over k,
(N2) A is a finitely generated $k[x_1, \ldots, x_r]$-module.

If $(x_1, , \ldots, x_r)$ is a Noether system of parameters of A, the polynomial algebra $P := k[x_1, , \ldots, x_r]$ is called a *Noether parameter algebra* of A.

We ask the reader to believe, to prove, or to check in the appropriate literature (see for example [Ben93], Sect. 4.3) the following fundamental result.

Theorem 7.1.26 *Let A be a finitely generated graded k-algebra.*

(1) *There exists a Noether system of parameters of A.*
(2) *All Noether systems of parameters of A have the same cardinal, equal to* Krdim(A).
(3) *The following assertions are equivalent.*

   (i) *There is a Noether system of parameters $(x_1, \ldots, x_r)$ of A such that A is a free module over $k[x_1, \ldots, x_r]$.*
   (ii) *Whenever $(x_1, \ldots, x_r)$ is a Noether system of parameters of A, A is a free module over $k[x_1, \ldots, x_r]$.*

Definition 7.1.27 When the equivalent assertions of item (3) of the above theorem are satisfied, we say that A is a *Cohen-Macaulay algebra*.

Proposition 7.1.28 *Assume that $(x_1, \ldots, x_m)$ is a system of homogeneous elements of A such that*

- $m \leq$ Krdim(A),
- A *is finitely generated as a $k[x_1, \ldots, x_m]$-module.*

Then

(1) $m =$ Krdim(A),
(2) $(x_1, \ldots, x_m)$ *is a system of Noether parameters of A.*

Brief sketch of proof Since an analogous result holds for field extensions (see (Tr1) and (TR2) above), it suffices to prove that the field of fractions of A is algebraic over the field of fractions of $k[x_1, \ldots, x_m]$. $\qquad\square$

Polynomial algebras and Noether parameters.

We shall now give some characterizations of systems of parameters of a polynomial algebra.

Proposition 7.1.29 *Let $S = k[v_1, \ldots, v_r]$ be a polynomial algebra, where $(v_1, \ldots, v_r)$ is a family of homogeneous algebraically independent elements with degrees $(e_1, \ldots, e_r)$.*

Let $(u_1, \ldots, u_r)$ be a family of nonconstant homogeneous elements of S with degrees respectively $(d_1, \ldots, d_r)$. We set $R := k[u_1, \ldots, u_r]$, and we denote by $\mathfrak{M}$ the maximal graded ideal of R.

(1) *The following assertions are equivalent.*

   (i)   *$S/\mathfrak{M}S$ is a finite dimensional k-vector space.*
   (ii) *S is a finitely generated R-module.*
   (iii) *$(u_1, \ldots, u_r)$ is a system of Noether parameters of S.*

(2) *If the preceding conditions hold, then S is a free R-module, and its rank is $\prod_i d_i / \prod_i e_i$.*

Proof Let us prove (1).

   (i)$\Rightarrow$(ii) results from Nakayama's Lemma (see 7.1.14, (4)).

   (ii)$\Rightarrow$(iii) results from Proposition 7.1.28.

   (iii)$\Rightarrow$(i) results from the general properties of systems of parameters (see Theorem 7.1.26, (3)).

Let us prove (2).

Since S is free over itself, it is Cohen-Macaulay (see Theorem 7.1.26, (4)), hence S is free over R. By Lemma 7.1.19, (1),

$$S \cong R \otimes_k (S/\mathfrak{M}S),$$

which implies

$$\mathrm{grdim}(S) = \mathrm{grdim}(R)\mathrm{grdim}(S/\mathfrak{M}S).$$

By Example 7.1.4, it follows that

$$\mathrm{grdim}(S/\mathfrak{M}S) = \frac{\prod_i(1 + q + \cdots + q^{d_i - 1})}{\prod_i(1 + q + \cdots + q^{e_i - 1})}$$

hence

$$[(S/\mathfrak{M}S) : K] = \mathrm{grdim}(S/\mathfrak{M}S)_{|q=1} = \frac{\prod_i d_i}{\prod_i e_i}.$$

$\square$

7.2 Graded Characters of Graded kG-modules

From now on, K denotes a characteristic zero field.

7.2.1 Notation and Definitions

- *A graded KG-module* is a graded K-vector space $M = \bigoplus_n M_n$ endowed with an operation of G (i.e., for each n, M_n is a KG-module).
- *The graded character* of a graded KG-module M is the class function

$$\mathrm{grchar}_M : G \to K((q))$$

defined by

$$\mathrm{grchar}_M(g) := \sum_n \mathrm{tr}(g, M_n) q^n \, .$$

In particular, $\mathrm{grchar}_M(1) = \mathrm{grdim}_K(M)$.

Example 7.2.1 For V a (finite K-dimensional) KG module, the symmetric algebra $S(V)$ and the exterior algebra $\Lambda(V)$ are naturally endowed with structures of graded KG-modules.

Lemma 7.2.2 *Let V be a KG-module, finite dimensional over K.*

(1) $\mathrm{grchar}_{\Lambda(V)}(g) = \det_V(1 + gq)$.

(2) $\mathrm{grchar}_{S(V)}(g) = \dfrac{1}{\det_V(1 - gq)}$.

(3) $\mathrm{grchar}_{\Lambda(V)}(-g)\mathrm{grchar}_{S(V)}(g) = 1$.

Proof For $g \in G$, let $(e_1, \ldots, e_r)$ be a basis of V consisting of eigenvectors of g, such that $ge_i = \lambda_i e_i$ for $i = 1, \ldots, r$. In what follows we use Proposition 7.1.13.

  (1) Given a natural integer m, we know that the family

$$(e_{i_1} \wedge \cdots \wedge e_{i_m})_{1 \le i_1 < \cdots < i_m \le r}$$

is a basis of $\Lambda(V)_m$.

  But for $1 \le i_1 < \cdots < i_m \le r$, we have

$$g(e_{i_1} \wedge \cdots \wedge e_{i_m}) = \lambda_{i_1} \cdots \lambda_{i_m}(e_{i_1} \wedge \cdots \wedge e_{i_m}) \, ,$$

which shows that

$$\mathrm{tr}(g, \Lambda(V)_m) = \sum_{i_1 < \cdots < i_m} \lambda_{i_1} \cdots \lambda_{i_m}$$

hence

$$\mathrm{grchar}_{\Lambda(V)}(g) = (1 + \lambda_1 q) \cdots (1 + \lambda_r q) \,,$$

and this proves (1).

(2) Similarly, given an integer m, we know that the family

$$(e_1^{d_1} \cdots e_r^{d_r})_{0 \leq d_i \,, \, d_1 + \cdots + d_r = m}$$

is a basis of $S(V)_m$. Since

$$g(e_1^{d_1} \cdots e_r^{d_r}) = \lambda_1^{d_1} \cdots \lambda_r^{d_r} (e_1^{d_1} \cdots e_r^{d_r}) \,,$$

it follows that

$$\mathrm{tr}(g, S(V)_m) = \sum_{d_1 + \cdots + d_r = m} \lambda_1^{d_1} \cdots \lambda_r^{d_r}$$

hence

$$\mathrm{grchar}_{S(V)}(g) = \frac{1}{(1 - \lambda_1 q) \cdots (1 - \lambda_r q)} \,,$$

and this proves (2).

(3) is immediate from (1) and (2). $\square$

Remark 7.2.3 Item (3) of the preceding lemma is a consequence of a much deeper result connecting $S(V)$ and $\Lambda(V)$, namely the exactness of the *Koszul complex* (see for example [Ben93, Sect. 4.2]).

Let S be an irreducible KG-module. We recall that we set $D_S = \mathrm{End}_{KG}(S)$, and that if χ denotes the character of S, we have

$$[D_S : K] = \langle \chi, \chi \rangle \,.$$

Definition 7.2.4 • For M a graded KG-module, *the graded multiplicity* of S in M is the formal series $\mathrm{grmult}(S, M) \in \mathbb{Z}((q))$ defined by

$$[D_S : K]\mathrm{grmult}(S, M) := \mathrm{grdim}_K \mathrm{Hom}_{KG}(S, M) \,.$$

• We call *S-isotypic component of M* and we denote by M_S the direct sum of all S-isotypic components of the homogeneous spaces M_n for $n \in \mathbb{Z}$. Thus, M_S is a graded KG-submodule of M.

Proposition 7.2.5 *Let M be a graded KG-module, and let S be an irreducible KG-module with character χ.*

(1) $\mathrm{grchar}_{M_S} = \mathrm{grmult}(S, M) \cdot \chi$, *that is, the product of* $\mathrm{grmult}(S, M) \in \mathbb{Z}((q))$ *by the class function* $\chi : G \to K$, *which is a class function* $G \to K((q))$.

(2) $\langle \chi, \chi \rangle \mathrm{grmult}(S, M) = \dfrac{1}{|G|} \sum_{g \in G} \mathrm{grchar}_M(g) \chi(g^{-1})$.

Proof (1) is clear. Let us prove (2). By definition,

$$\langle \chi, \chi \rangle \mathrm{grmult}(S, M) = \mathrm{grdim}_K \mathrm{Hom}_{KG}(S, M)$$

$$= \bigoplus_n \dim \mathrm{Hom}_{KG}(S, M_n) q^n$$

$$= \sum_n \langle \chi_{M_n}, \chi \rangle q^n = \sum_n \frac{1}{|G|} \sum_{g \in G} \chi_{M_n}(g) \chi(g^{-1}) q^n$$

$$= \frac{1}{|G|} \sum_{g \in G} \left(\sum_n \chi_{M_n}(g) q^n \right) \chi(g^{-1})$$

$$= \frac{1}{|G|} \sum_{g \in G} \mathrm{grchar}_M(g) \chi(g^{-1}) .$$

$\square$

We also set (using preceding notation):

$$\mathrm{grmult}(\chi, M) := \mathrm{grmult}(S, M) ,$$

and for any $\alpha \in \mathrm{CF}(G, \mathbb{Q}_K^G((q)))$ and $\beta \in \mathrm{CF}(G, \mathbb{Q}_K^G)$, we set

$$\langle \alpha, \beta \rangle := \frac{1}{|G|} \sum_{g \in G} \alpha(g) \beta(g)^* .$$

7.2.2 *Isotypic Components of the Symmetric Algebra*

Preliminary.
Let B be an integral domain, with field of fractions L. Let G be a finite group of automorphisms of B. We set $A := \mathrm{Fix}^G(B)$, the subring of G-fixed points of B, and we denote by K its field of fractions.

Proposition 7.2.6 *In the above situation, we have:*

(1) *B is integral over A.*

(2) *Any element of L can be written* $\dfrac{b}{a}$ *with* $a \in A$ *and* $b \in B$. *We have* $K = \mathrm{Fix}^G(L)$ *and* L/K *is a Galois extension, with G as Galois group.*

(3) *If B is integrally closed, A is also integrally closed.*

Proof (1) Every $b \in B$ is a root of the polynomial $P_b(t) := \prod_{g \in G}(t - g(b))$, which belongs to $A[t]$.

(2) For $b_1, b_2 \in B$ and $b_2 \neq 0$, we have

$$\frac{b_1}{b_2} = \frac{b_1 \prod_{g \in G, g \neq 1} g(b_2)}{\prod_{g \in G} g(b_2)},$$

which proves the first assertion of (2). This makes the second assertion clear.

(3) An element of K which is integral over A is *a fortiori* integral over B, whence it belongs to B and so to $B \cap K$. But $B \cap K = B \cap L^G = B^G = A$. $\square$

Fixed points on the symmetric algebra.

From now on, we let $S(V)$ be the symmetric algebra of an r-dimensional vector space V over the field K. We let $\mathfrak{N}$ be its graded maximal ideal, so that $S(V)/\mathfrak{N} = K$. Let $\mathcal{L}$ be the field of fractions of $S(V)$. Notice that $V = \mathfrak{N}/\mathfrak{N}^2$.

Let G be a finite group of automorphisms of V.

We denote by $S(V)^G$ the subring of fixed points of G on $S(V)$ and we set $\mathfrak{M} := S(V)^G \cap \mathfrak{N} = \mathfrak{N}^G$, the maximal graded ideal of $S(V)^G$. Let $\mathcal{K}$ be the field of fractions of $S(V)^G$.

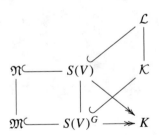

Lemma 7.2.7 (1) $S(V)$ *is a finitely generated* $S(V)^G$*-module.*

(2) $S(V)^G$ *is a finitely generated* K*-algebra.*

Proof (1) Since $S(V)$ is a K-algebra of finite type, $S(V)$ is a *fortiori* of finite type over $S(V)^G$. Since $S(V)$ is integral over $S(V)^G$, $S(V)$ is then a finitely generated $S(V)^G$-module (see Appendix C).

(2) Assume $S(V) = K[x_1, \ldots, x_r]$. Let $P_1(t), \ldots, P_r(t) \in S(V)^G[t]$ be nonzero polynomials having respectively $x_1, \ldots, x_r$ as roots. Let $C_1, \ldots, C_r$ denote respectively the set of coefficients of $P_1(t), \ldots, P_r(t)$. Thus, $S(V)$ is integral over $K[C_1, \ldots, C_r]$, and since $S(V)$ is a finitely generated algebra over $K[C_1, \ldots, C_r]$,

$S(V)$ is a finitely generated module over the algebra $K[C_1, \ldots, C_r]$. Since $K[C_1, \ldots, C_r]$ is noetherian, the submodule $S(V)^G$ of $S(V)$ is also finitely generated (see Appendix D). Hence $S(V)^G$, a finitely generated module over a finitely generated K-algebra, is a finitely generated K-algebra. $\qquad\square$

Isotypic components of the symmetric algebra.

The algebra $S(V)$ is a graded KG-module. For all irreducible character χ of G on K, we let $S(V)_\chi$ denote the χ-isotypic component of $S(V)$. Note that if 1_G is the trivial character of G, then $S(V)_{1_G}$ is the subalgebra $S(V)^G$ of fixed points of $S(V)$ under G.

Lemma 7.2.8 (1) *Each $S(V)_\chi$ is a graded KG-module, and a graded $S(V)^G$-submodule of $S(V)$.*

(2) *We have*

$$S(V) = \bigoplus_{\chi \in \mathrm{Irr}_K(G)} S(V)_\chi .$$

(3) *For χ the character of an irreducible KG-module we have (with obvious notation)*

$$\langle \chi, \chi \rangle \mathrm{grmult}(\chi, S(V)) = \frac{1}{|G|} \sum_{g \in G} \frac{\chi(g^{-1})}{\det_V(1 - gq)} ,$$

$$\mathrm{grchar}_{S(V)_\chi} = \mathrm{grmult}(\chi, S(V)) \cdot \chi .$$

(4) *[MOLIEN'S FORMULA] In particular,*

$$\mathrm{grdim}_K S(V)^G = \frac{1}{|G|} \sum_{g \in G} \frac{1}{\det_V(1 - gq)} .$$

Proof (1) Whenever $S(V)_n$ is a homogeneous component of degree n of $S(V)$, multiplication by a homogeneous element $x \in S(V)^G$ defines an isomorphism of KG-modules from $S(V)_n$ to $xS(V)_n$. Thus multiplication by x sends $S(V)_\chi$ into itself, which proves (1).

(2) is immediate. (3) and (4) results from Proposition 7.2.5. $\qquad\square$

Proposition 7.2.9 *Whenever P is a Noether parameter algebra of $S(V)^G$, and whenever $\chi \in \mathrm{Irr}_K(G)$, the isotypic component $S(V)_\chi$ is free over P.*

In particular the invariant algebra $S(V)^G$ is Cohen–Macaulay.

Proof A Noether parameter algebra P of $S(V)^G$ is a Noether parameter algebra of $S(V)$. Since $S(V)$ is free over itself, $S(V)$ is free over P by item (3) of Theorem 7.1.26 (see also Definition 7.1.27). The proposition follows from Lemma 7.2.8, (2). $\qquad\square$

7.2.3 Computations with Power Series

Let P be a Noether parameter algebra of $S(V)^G$. We denote by $\mathfrak{M}_P$ the unique maximal graded ideal of P.

Let m denote the rank of $S(V)^G$ over P, which is also the dimension of the graded K-vector space $S(V)^G/\mathfrak{M}_P S(V)^G = K \otimes_P S(V)^G$.

We call P-*exponents* of $S(V)^G$ the family $(e_1, e_2, \ldots, e_m)$ of integers such that

$$\operatorname{grdim}_K(S(V)^G/\mathfrak{M}_P S(V)^G) = q^{e_1} + q^{e_2} + \cdots + q^{e_m} .$$

Thus there exists a K-basis of $S(V)^G/\mathfrak{M}_P S(V)^G$ consisting of homogeneous elements of degrees respectively $e_1, e_2, \ldots, e_m$.

Let $d_1, d_2, \ldots, d_r$ be the characteristic degrees of P (see Definition 7.1.23), so that

$$\operatorname{grdim}_K(P) = \frac{1}{(1 - q^{d_1})(1 - q^{d_2}) \cdots (1 - q^{d_r})} .$$

Since $S(V)^G$ is free over P, we have

$$S(V)^G \cong P \otimes_K (S(V)^G/\mathfrak{M}_P S(V)^G) .$$

Thus the graded dimension of $S(V)^G$ is

$$\operatorname{grdim}_K(S(V)^G) = \frac{q^{e_1} + q^{e_2} + \cdots + q^{e_m}}{(1 - q^{d_1})(1 - q^{d_2}) \cdots (1 - q^{d_r})} .$$

Now let $\chi \in \operatorname{Irr}_K(G)$. Let $\chi(1)m_\chi$ denote the rank of $S(V)_\chi$ over P, which is equal to the dimension of the graded K-vector space

$$S(V)_\chi/\mathfrak{M}_P S(V)_\chi = K \otimes_P S(V)_\chi .$$

Since each homogeneous component of $S(V)_\chi/\mathfrak{M}_P S(V)_\chi$ is a direct sum of modules with character χ, the graded dimension of the graded vector space $S(V)_\chi/\mathfrak{M}_P S(V)_\chi$ has the form

$$\operatorname{grdim} S(V)_\chi/\mathfrak{M}_P S(V)_\chi = \chi(1)(q^{e_1(\chi)} + q^{e_2(\chi)} + \cdots + q^{e_{m_\chi}(\chi)}) ,$$

for suitable $e_i(\chi) \in \mathbb{N}$, from which we deduce

$$\operatorname{grdim}_K(S(V)_\chi) = \chi(1) \frac{q^{e_1(\chi)} + q^{e_2(\chi)} + \cdots + q^{e_{m_\chi}(\chi)}}{(1 - q^{d_1})(1 - q^{d_2}) \cdots (1 - q^{d_r})} .$$

Thus

$$\text{grmult}(\chi, S(V)) = \frac{q^{e_1(\chi)} + q^{e_2(\chi)} + \cdots + q^{e_{m_\chi}(\chi)}}{(1 - q^{d_1})(1 - q^{d_2}) \cdots (1 - q^{d_r})}. \tag{7.2.10}$$

Theorem 7.2.11 *Let K be any characteristic zero field, let V be a finite dimensional K-vector space, and let G be a finite subgroup of $\mathrm{GL}(V)$.*

Let P be a parameter algebra for the algebra of invariants $S(V)^G$, with degrees $(d_i)_{1 \le i \le r}$. Let us denote by m the rank of $S(V)^G$ over P.

(1) $|G|m = \prod_{i=1}^{r} d_i$.
(2) *As KG-modules, the ungraded module $S(V)/\mathfrak{M}_P S(V)$ is isomorphic to $(KG)^m$.*
(3) *The (ungraded) PG-module $S(V)$ is isomorphic to $(PG)^m$.*

Proof (1) We first prove the following lemma.

Lemma 7.2.12 *Let $r, n, d_1, d_2, \ldots, d_r, e_1, e_2, \ldots, e_i$ be positive integers, and let*

$$\alpha(q) := \frac{q^{e_1} + q^{e_2} + \cdots + q^{e_i}}{(1 - q^{d_1})(1 - q^{d_2}) \cdots (1 - q^{d_r})} \in \mathbb{Q}(q).$$

Let

$$\alpha(q) = \frac{a_r}{(1 - q)^r} + \frac{a_{r-1}}{(1 - q)^{r-1}} + \cdots$$

be the Laurent expansion of $\alpha(q)$ around $q = 1$. Then

$$a_r (d_1 d_2 \cdots d_r) = n .$$

Proof of Lemma 7.2.12 Set

$$\alpha_1(q) := (1 - q)^r \alpha(q) .$$

Then

$$\alpha_1(q) = \frac{q^{e_1} + q^{e_2} + \cdots + q^{e_i}}{(1 + q + \cdots + q^{d_1 - 1}) \cdots (1 + q + \cdots + q^{d_r - 1})} .$$

Then we get the announced equality by evaluating $\alpha_1(q)$ at $q = 1$.   □

Let us notice the following particular case of what precedes. Let P be a polynomial algebra with characteristic degrees $d_1, d_2, \ldots, d_r$. Then

$$\text{grdim}_K P \in \frac{1}{d_1 d_2 \cdots d_r} \frac{1}{(1 - q)^r} + \frac{1}{(1 - q)^{r-1}} K[q] . \tag{7.2.13}$$

We remark that by Eq. 7.2.10, by item (3) of Lemma 7.2.8, and since G acts faithfully on V,

$$\langle \chi, \chi \rangle \frac{q^{e_1(\chi)} + \cdots + q^{e_{m_\chi}(\chi)}}{(1 - q^{d_1}) \cdots (1 - q^{d_r})} \in \chi(1) \frac{1}{|G|} \frac{1}{(q - 1)^r} + \frac{1}{(q - 1)^{r-1}} K[q],$$

from which, applying the preceding lemma, we get

$$\chi(1) \frac{1}{|G|} d_1 d_2 \cdots d_r = \langle \chi, \chi \rangle m_\chi. \tag{7.2.14}$$

Specializing the above formulae to the case where $\chi = 1_G$, we get

$$\frac{1}{|G|} d_1 d_2 \cdots d_r = m.$$

which is item (1) of Theorem 7.2.11.

(2) Substituting (1) in the equality (7.2.14) gives

$$\chi(1)m = \langle \chi, \chi \rangle m_\chi. \tag{7.2.15}$$

In order to prove the second assertion of Theorem 7.2.11 it suffices to show that the character of the (ungraded) KG-module $S(V)/\mathfrak{M}_P S(V)$ equals the character of $(KG)^m$.

By item (1) of Proposition 3.2.26, we know that the character of $(KG)^m$ equals $\sum_\chi m(\chi(1)/\langle \chi, \chi \rangle)\chi$, which, by (7.2.15), is $\sum_\chi m_\chi \chi$, that is, the character of $S(V)/\mathfrak{M}_P S(V)$.

(3) By the first item of Lemma 7.1.19, since $S(V)$ is a free graded P-module, we have an isomorphism of graded P-modules

$$P \otimes_K (S(V)/\mathfrak{M}_P S(V)) \xrightarrow{\sim} S(V).$$

This implies that $S(V) \cong P \otimes_K (KG)^m \cong (PG)^m$.   □

7.2.4   A Simple Example

Let us consider $V := K^2$ and $G := \left\{ \begin{pmatrix} 1 & 0 \\ 0 & 1 \end{pmatrix}, \begin{pmatrix} -1 & 0 \\ 0 & -1 \end{pmatrix} \right\} \subset GL_2(K)$. As before, we set $S(V) := K[x, y]$.

By Molien's formula (Lemma 7.2.8), one can check that

$$\mathrm{grdim}(S(V)^G) = \sum_{n=0}^{\infty} (2n + 1)q^n = \frac{1 - q^4}{(1 - q^2)^3} = \frac{1 + q^2}{(1 - q^2)^2}.$$

Since $x^2, xy, y^2 \in K[x, y]^G$, we have

$$K[x^2, xy, y^2] \quad \text{and} \quad K[x^2, y^2] + K[x^2, y^2]xy \subseteq K[x, y]^G.$$

It is easy to check that

$$\begin{cases} \operatorname{grdim}(K[x^2, xy, y^2]) = \dfrac{1 - q^4}{(1 - q^2)^3}, \\[3mm] \operatorname{grdim}(K[x^2, y^2] \oplus K[x^2, y^2]xy) = \dfrac{1 + q^2}{(1 - q^2)^2}, \end{cases}$$

from which one deduces

$$\begin{cases} S(V)^G = K[x^2, xy, y^2] \cong K[u, v, w]/(uw - v^2), \\[2mm] S(V)^G = K[x^2, y^2] \oplus K[x^2, y^2]xy. \end{cases}$$

7.2.5 Complement: Reflection Groups

One can prove the following theorem (see for example [Bou68, Sect. 5] or [Bro10, Chap. 4]), due to Shephard–Todd (using the classification of irreducible complex reflection groups) and to Chevalley and Serre for a classification-free proof.

Remark 7.2.16 This theorem generalizes the well known following description of symmetric polynomials.

Let us denote by

$$\Sigma_1(x_1, \ldots, x_r) := x_1 + \cdots + x_r,$$

$$\Sigma_2(x_1, \ldots, x_r) := \sum_{1 \le i < j \le r} x_i x_j,$$

$$\vdots$$

$$\Sigma_r(x_1, \ldots, x_r) := x_1 \cdots x_r,$$

the family of elementary symmetric polynomials. Then the subalgebra of $K[x_1, \ldots, x_r]$ of all symmetric polynomials is $K[\Sigma_1, \ldots, \Sigma_r]$.

Theorem 7.2.17 *Let K be a characteristic zero field, let V be a finite dimensional K-vector space. Let G be a finite subgroup of $\operatorname{GL}(V)$.*

(1) *The following assertions are equivalent.*

    (i)   *G is generated by reflections.*
    (ii) *$S(V)^G$ is a polynomial algebra.*

(2) *If this is the case, let us denote by $(d_1, d_2, \ldots, d_r)$ the characteristic degrees of $S(V)^G$. Then*

(a) $|G| = d_1 d_2 \cdots d_r$,
(b) *As ungraded* $S(V)^G G$*–modules, we have* $S(V) \cong S(V)^G G$.

Proof Here we only prove (2). Since $S(V)^G$ is a polynomial algebra, it is a Noether parameter algebra for itself, that is, we may choose $P = S(V)^G$ in Theorem 7.2.11. Then $m = 1$, and the assertion is a particular case of Theorem 7.2.11. $\square$

Exercise 7.2.18 Let $(x_1, x_2, \ldots, x_r)$ be a basis of V. As above, we denote by $(\Sigma_j(x_1, x_2, \ldots, x_r))_{1 \le j \le r}$ the family of elementary symmetric polynomials.

Let $d \ge 1$ and $e \ge 1$ be integers. We refer to Sect. 4.2.12 for the definition of the group $G(de, e, r)$. Let us set

$$\begin{cases} f_j := \Sigma_j(x_1^{de}, x_2^{de}, \ldots, x_r^{de}) & \text{for } 1 \le j \le r - 1, \\ f_r := (x_1 x_2 \cdots x_r)^d . \end{cases}$$

Prove that

$$\mathbb{C}[x_1, x_2, \ldots, x_r]^{G(de,e,r)} = \mathbb{C}[f_1, f_2, \ldots, f_r] .$$

More Exercises on Chap. 7

In the following exercises, K is a characteristic zero field, V is an r-dimensional K-vector space, and G is a finite subgroup of $GL(V)$. Let $\mathrm{Ref}(G)$ be the set of reflections of G (we recall that here a reflection is not necessarily of order 2).

Exercise 7.2.19 (1) For all hyperplanes H of V, we denote by $C_G(H)$ the subgroup of G consisting of the elements of G which fix H pointwise. Prove that $C_G(H)$ is a cyclic subgroup of G, whose non trivial elements are reflections.

(2) Let us denote by $\mathcal{A}(G)$ the set of reflecting hyperplanes of G, that is, the set of hyperplanes H such that $C_G(H) \neq 1$. Prove that

$$\mathrm{Ref}(G) = \bigsqcup_{H \in \mathcal{A}(G)} (C_G(H) \backslash \{1\}) .$$

Exercise 7.2.20 (1) Prove that

$$|\mathrm{Ref}(G)| = 2 \sum_{s \in \mathrm{Ref}(G)} \frac{1}{1 - \det(s)} .$$

(2) From Molien's formula (see Lemma 7.2.8), prove that the Laurent series expansion of $\mathrm{grdim}_K S(V)^G$ around $q = 1$ is

$$\mathrm{grdim}_K S(V)^G = \frac{1}{|G|} \left(\frac{1}{(q-1)^r} + \frac{|\mathrm{Ref}(G)|}{2} \frac{1}{(q-1)^{r-1}} + \cdots \right) .$$

Exercise 7.2.21 Now we assume that G is generated by $\mathrm{Ref}(G)$. Hence (see Theorem 7.2.17) $S(V)^G$ is a polynomial algebra. Let us denote by $(d_1, \dots, d_r)$ the family of characteristic degrees of $S(V)^G$, so that

$$\mathrm{grdim}_K S(V)^G = \frac{1}{(1 - q^{d_1}) \cdots (1 - q^{d_r})} .$$

(1) Prove that the Laurent series expansion of $\mathrm{grdim}_K S(V)^G$ in $q - 1$ is

$$\mathrm{grdim}_K S(V)^G = \frac{1}{d_1 \cdots d_r} \left(\frac{1}{(q-1)^r} + \frac{\sum_{j=1}^r (d_j - 1)}{2} \frac{1}{(q-1)^{r-1}} + \cdots \right) .$$

(2) From Exercise 7.2.20, deduce that

$$|\mathrm{Ref}(G)| = \sum_{j=1}^r (d_j - 1) .$$

Chapter 8
Drinfeld Double

Let k be a field.

In this chapter, all vector spaces and all tensor products will be over k. Hence the letter k will be omitted: we shall write $x \otimes y$ instead of $x \otimes_k y$, and for a k-vector space V, we shall write $\dim V$ instead of $[V:k]$.

8.1 The Drinfeld Double of a Finite Group as an Algebra

The following definition is a particular case of a general construction due to Drinfeld [Dri87], which applies to any Hopf k-algebra (see below Sect. 7.2) and its k-dual – here we only deal with the particular case of the Hopf algebra kG (G a finite group) and its dual $\mathrm{F}(G, k)$. It consists, in particular, of endowing $D_k G = \mathrm{F}(G, k) \rtimes G$ with a structure of algebra.

Vladimir Drinfeld (born 1954)

© Springer Nature Singapore Pte Ltd. 2017
M. Broué, *On Characters of Finite Groups*, Mathematical Lectures from Peking University, https://doi.org/10.1007/978-981-10-6878-2_8

8.1.1   The Semidirect Product of kG and Its Dual

Let G be a finite group. We recall that kG denotes the group algebra of G over k.

- The dual $\mathrm{Hom}_k(kG, k)$ of kG (as a k-vector space) is identified with $\mathrm{F}(G, k)$, the algebra of functions from G to k: if $(g)_{g \in G}$ is the natural basis of kG, its dual basis is the family $(\delta_s)_{s \in G}$ in $\mathrm{F}(G, k)$ where $\delta_s(g) = 0$ if $g \neq s$ and $\delta_s(s) = 1$.
- The space $\mathrm{F}(G, k)$ is endowed (by pointwise multiplication) with a natural structure of commutative k-algebra, where the family $(\delta_s)_{s \in G}$ is a k-basis consisting of idempotents. Thus

$$\mathrm{F}(G, k) = \bigoplus_{s \in G} \mathrm{F}(G, k)\delta_s \quad \text{where} \quad \mathrm{F}(G, k)\delta_s \simeq k.$$

The unit element of $\mathrm{F}(G, k)$ is $\sum_{s \in G} \delta_s$.
- The group G acts on $\mathrm{F}(G, k)$: for $g \in G$ and $\varphi \in \mathrm{F}(G, k)$, we set ${}^g\varphi := g\varphi g^{-1}$, that is $({}^g\varphi)(s) := \varphi(g^{-1}sg)$, for all $s \in G$. Thus we may define the *semidirect product*

$$D_k G := \mathrm{F}(G, k) \rtimes G$$

of $\mathrm{F}(G, k)$ with kG as the free $\mathrm{F}(G, k)$-module with basis $(g)_{g \in G}$, and product defined by

$$\varphi.g.\varphi'.g' := \varphi.{}^g\varphi'.gg' \quad \text{where} \quad \varphi.{}^g\varphi' \in \mathrm{F}(G, k) \text{ and } gg' \in G.$$

In other words, we have $D_k G = \mathrm{F}(G, k) \otimes kG$, and

$$(\varphi \otimes g) \cdot (\varphi' \otimes g') = \varphi({}^g\varphi') \otimes gg'.$$

In particular, for $s, t, g, h \in G$, $(\delta_s \otimes g)(\delta_t \otimes h) = 0$ unless $s = gtg^{-1}$, in which case $(\delta_s \otimes g)(\delta_{g^{-1}sg} \otimes h) = \delta_s \otimes gh$.

The algebra $D_k G$ is called the *Drinfeld double of G over k*.

Note that the unit element 1_D of $D_k G$ is $1_D := \sum_{s \in G} \delta_s \otimes 1$.

From now on, we identify $\mathrm{F}(G, k)$ with the subspace $\mathrm{F}(G, k) \otimes 1$ of $D_k G$, and the group algebra kG with the subspace $1 \otimes kG$ of $D_k G$.

Thus we have the following rules for the algebra $D_k G$.

- $D_k G$ has for k-basis the family $(\delta_s g)_{s, g \in G}$.
- $D_k G = \bigoplus_{g \in G} \mathrm{F}(G, k)g$.
- For $\varphi \in \mathrm{F}(G, k)$ and $g \in G$,

$$g\varphi = {}^g\varphi g \quad \text{where} \quad ({}^g\varphi)(s) := \varphi(g^{-1}sg).$$

In particular (denoting by $\delta_{x,y}$ the Kronecker symbol),

$$\delta_s g \delta_t h = \delta_s \delta_{gtg^{-1}} gh = \boldsymbol{\delta}_{s,gtg^{-1}} \delta_s gh \quad \text{and} \quad g\delta_s = \delta_{gsg^{-1}} g \,.$$

- $1_D = \sum_{s \in G} \delta_s.$

Center of $D_k G$.

Notation 8.1.1 The group G acts by simultaneous conjugation on the set of commuting pairs

$$(G \times G)^{\mathrm{com}} := \{(g, h) \mid g, h \in G, \ gh = hg\} \,,$$

and we denote by $\mathrm{Cl}((G \times G)^{\mathrm{com}})$ the set of orbits.

The projection $G \times G \to G$ onto the first component induces a bijection between $\mathrm{Cl}((G \times G)^{\mathrm{com}})$ and the set of G-conjugacy classes of pairs (g, D) where $g \in G$ and D runs over $\mathrm{Cl}(C_G(g))$.

Let $x = \sum_{s,g \in G} \lambda_{s,g} \delta_s g \in D_k G$ where $\lambda_{s,g} \in k$. Then $x \in Z(D_k G)$ if and only if the following two conditions are satisfied:

(1) For all $t \in G$, $\delta_t x = x \delta_t$, which is equivalent to

$$(\forall t \in G), \ \sum_{g \in G} \lambda_{t,g} \delta_t g = \sum_{g \in G} \lambda_{t,g} \delta_{gtg^{-1}} g \,,$$

that is, since $(g)_{g \in G}$ is a basis of $D_k G$ over $F(G, k)$, for all $s, g \in G$,

$$\lambda_{s,g} \neq 0 \Rightarrow s = gsg^{-1} \ i.e., \ (s, g) \in (G \times G)^{\mathrm{com}} \,.$$

In other words, the subalgebra of $D_k G$ which commutes with all the δ_t for $t \in G$ has for basis $(\delta_s g)$ where (s, g) runs over the set $(G \times G)^{\mathrm{com}}$ of commuting pairs of $G \times G$.

(2) If the preceding condition is satisfied, then for all $h \in G$, $hxh^{-1} = x$, which is equivalent to

$$\sum_{(s,g) \in (G \times G)^{\mathrm{com}}} \lambda_{s,g} \delta_s g = \sum_{(s,g) \in (G \times G)^{\mathrm{com}}} \lambda_{s,g} \delta_{hsh^{-1}} hgh^{-1} \,,$$

that is, $\lambda_{s,g}$ is constant on the G-orbits on $(G \times G)^{\mathrm{com}}$.

For $C \in \mathrm{Cl}((G \times G)^{\mathrm{com}})$, let us set

$$\Sigma_C := \sum_{(s,g) \in C} \delta_s g \,.$$

In other words, any $x \in Z(D_k G)$ may be written

$$x = \sum_{C \in \mathrm{Cl}((G \times G)^{\mathrm{com}})} \lambda_C \Sigma_C \, , \text{ or}$$

$$x = \sum_{s \in [\mathrm{Cl}(G)]} \sum_{r \in [G/C_G(s)]} r \cdot \left(\delta_s \sum_{D \in \mathrm{Cl}(C_G(s))} \lambda_{s,D} \mathcal{S} D \right) \cdot r^{-1} \, ,$$

and this proves the following proposition.

Proposition 8.1.2

(1) The family $(\Sigma_C)_{C \in \mathrm{Cl}((G \times G)^{\mathrm{com}})}$ is a k-basis of $Z(D_k G)$.
(2) The map

$$\prod_{s \in [\mathrm{Cl}(G)]} Z k C_G(s) \to Z(D_k G) \, ,$$

$$(z_s)_{s \in [\mathrm{Cl}(G)]} \mapsto \sum_{s \in [\mathrm{Cl}(G)]} \sum_{r \in [G/C_G(s)]} r(\delta_s z_s) r^{-1} \, ,$$

is an isomorphism of k-vector spaces.

Exercise 8.1.3 Prove that the k-linear isomorphism of item (2) in the above Proposition 8.1.2 is an *algebra isomorphism*.

$D_k G$ as a symmetric algebra.

Let us consider the following k-linear map

$$\tau : \delta_s g \mapsto \delta_{g,1}$$

(recall $\delta_{x,y}$ denotes the Kronecker symbol). Then τ is a symmetrizing form (see Lemma 2.3.13) on $D_k G$. Indeed:

(1) τ is *central*, namely, for all $a, b \in D_k G$, $\tau(ab) = \tau(ba)$, since

$$\tau(\delta_s g \delta_t h) = \delta_{s, gtg^{-1}} \tau(\delta_s gh) = \delta_{s, gtg^{-1}} \delta_{gh,1} \, ,$$
$$\tau(\delta_t h \delta_s g) = \delta_{t, hsh^{-1}} \tau(\delta_t hg) = \delta_{t, hsh^{-1}} \delta_{hg,1} \, ,$$

which shows that $\tau(\delta_s g \delta_t h) = \tau(\delta_t h \delta_s g)$, and
(2) the map

$$\widehat{\tau} : D_k G \to \mathrm{Hom}_k(D_k G, k) \, , \quad a \mapsto \widehat{\tau}(a) : (x \mapsto \tau(ax))$$

is a k-linear isomorphism, since the above computation of $\tau(\delta_s g \delta_t h)$ also shows that the basis $(\delta_s g)_{s,g \in G}$ has as dual basis $(\delta_s g^{-1})_{s,g \in G}$.

As for the group algebra, we call *central functions* on $D_k G$ the k-linear forms $f : D_k G \to k$ such that, for all $a, b \in D_k G$, $f(ab) = f(ba)$. We denote by $\mathrm{CF}(D_k G, k)$ the space of central functions, and the isomorphism $\widehat{\tau}$ induces a k-linear isomorphism

$$\mathrm{CF}(D_k G, k) \xrightarrow{\sim} Z(D_k G) \ , \quad f \mapsto f^0 := \sum_{s,g \in G} f(\delta_s g^{-1}) \delta_s g \ .$$

The *regular representation* of $D_k G$ consists, as usual, in letting $D_k G$ act on itself by left multiplication. Let us denote by $\chi^{\mathrm{reg}}_{D_k G}$ the character of that representation, that is, the central function on $D_k G$ such that, for all $a \in D_k G$, $\chi^{\mathrm{reg}}_{D_k G}(a)$ is the trace of the endomorphism $x \mapsto ax$ of $D_k G$.

It is immediate to check that

$$\forall s, g \in G \ , \ \chi^{\mathrm{reg}}_{D_k G}(\delta_s g) = |G| \delta_{g,1} \ ,$$

which proves the next lemma.

Lemma 8.1.4 *We have*

$$\chi^{\mathrm{reg}}_{D_k G} = |G|\tau \quad and \quad (\chi^{\mathrm{reg}}_{D_k G})^0 = |G| 1_D \ .$$

8.1.2   A Description of $_{D_k G}\mathbf{mod}$

For an alternative approach to $_{D_k G}\mathbf{mod}$, one may read Lusztig [Lus87, Sect. 2].

George Lusztig (born 1946)

Let X be a $D_k G$-module. Since $D_k G = \bigoplus_{s \in G} \delta_s k G$ where $(\delta_s)_{s \in G}$ is a family of orthogonal idempotents with sum 1, we have

$$X = \bigoplus_{s \in G} \delta_s X \ ,$$

a decomposition stable under the action of G since

$$g(\delta_s X) = \delta_{g_s} X \,.$$

Notice that the centraliser $C_G(s)$ acts on $\delta_s X$.

Let $_{D_k G}\mathbf{mod}$ denote the category of $D_k G$-modules. By what precedes, it is easy to see that this category may be described as follows.

Objects and Morphisms

- The objects are G-graded kG-modules, i.e., kG-modules X endowed with a decomposition

$$X = \bigoplus_{s \in G} {}_s X \,,$$

  such that, for $g \in G$, $g(_s X) = {}_{g_s} X \,.$
- A morphism

$$f : X = \bigoplus_{s \in G} {}_s X \to Y = \bigoplus_{s \in G} {}_s Y$$

  between two objects is a kG-linear map such that, for all $s \in G$, $f(_s X) \subset {}_s Y$.

A category equivalent to $_{D_k G}\mathbf{mod}$ as an abelian category.

Let $[\mathrm{Cl}(G)]$ be a complete set of representatives of the conjugacy classes of G.

Proposition 8.1.5 *Let $_{D_k G}\mathbf{mod}$ be the category of $D_k G$-modules. As abelian categories,*

$$_{D_k G}\mathbf{mod} \simeq \bigoplus_{s \in [\mathrm{Cl}(G)]} {}_{kC_G(s)}\mathbf{mod} \,.$$

Proof Let us describe two inverse functors between these categories.

- To an object

$$X = \bigoplus_{s \in G} {}_s X$$

  of $_{D_k G}\mathbf{mod}$ one associates the object

$$\bigoplus_{s \in [\mathrm{Cl}(G)]} {}_s X \,,$$

  to a morphism $f : X \to Y$, one associates the family $(f_{|_s X})_{s \in [\mathrm{Cl}(G)]}$, which is indeed a morphism

$$\bigoplus_{s\in[\mathrm{Cl}(G)]} {}_sX \to \bigoplus_{s\in[\mathrm{Cl}(G)]} {}_sY .$$

- Let $s \in G$ and let V be a $kC_G(s)$-module. Consider the kG-module

$$X(s, V) := \mathrm{Ind}_{C_G(s)}^G V = \bigoplus_{r\in[G/C_G(s)]} (r \otimes_{kC_G(s)} V) .$$

We set

$$\begin{cases} {}_tX(s, V) := r \otimes_{kC_G(s)} V & \text{if } t = rsr^{-1} , \\ {}_tX(s, V) := 0 & \text{if } t \text{ is not conjugate to } s , \end{cases}$$

and we see that $X(s, V)$ is endowed with a G-grading which is stable under the action of G, that is, $X(s, V)$ is a D_kG-module.

Furthermore, to a morphism $f : V \to V'$ of $kC_G(s)$-modules, we associate the morphism

$$\mathrm{Ind}_{C_G(s)}^G(f) : X(s, V) \to X(s, V') ,$$

which is indeed a kG-morphism of graded G-modules. $\qquad\qquad\square$

Corollary 8.1.6

(1) The map $(s, S) \mapsto X(s, S)$ defines a bijection between

- *the family of pairs (s, S) where s runs over a complete set of representatives of conjugacy classes of G, and $S \in \mathrm{Irr}_k(C_G(s))$ (resp. S runs over a complete set of representatives of indecomposable $k(C_G(s))$-modules),*
- *a complete set of representatives $\mathrm{Irr}(D_kG)$ of irreducible D_kG-modules (resp. a complete set of representatives of indecomposable D_kG-modules).*

(2) Every D_kG-module is semisimple if and only if the characteristic of k does not divide $|G|$.

Proof (1) is an immediate consequence of Proposition 8.1.5. Let us prove (2). By Proposition 8.1.5, we see that every D_kG-module is semisimple if and only if for all $s \in G$ every $kC_G(s)$-module is semisimple, hence (see Remark 3.1.4) if and only if the characteristic of k does not divide $|G|$. $\qquad\qquad\square$

Description of $X(s, S)$.

By definition, $X(s, S)$ is "supported" by the conjugacy class of s, *i.e.*,

$$X(s, S) = \bigoplus_{r\in[G/C_G(s)]} {}_{rsr^{-1}}X(s, S) .$$

Moreover

$$_sX(s, S) = S \text{ as } kC_G(s) - \text{modules},$$

and more generally $_{rsr^{-1}}X(s, S) = {}^rS$ as $kC_G(rsr^{-1})$ − modules.

Notice in particular that

$$\dim X(s, S) = |G : C_G(s)| \dim S.$$

8.1.3   On the Center $Z(D_kG)$ and Central Functions Again

Character of $X(s, S)$.

Let us compute the trace of an element $\delta_t g$ on $X(s, S)$. For all $h \in G$,

$$\delta_t g({}^hS) = \delta_t {}^{gh}S = \delta_t \delta_{ghsh^{-1}g^{-1}} X(s, S) = \delta_{t,ghsh^{-1}g^{-1}} {}^{gh}S.$$

Since the subspace $_{rsr^{-1}}X(s, S) = {}^rS$ is stable under $\delta_t g$ if and only if $t = grsr^{-1}g^{-1}$, we see that

$$\text{tr}(\delta_t g; X(s, S)) = \begin{cases} 0 \text{ if } t \notin \text{Cl}(s), \\ 0 \text{ if } t = rsr^{-1} \text{ but } r^{-1}gr \notin C_G(s), \\ \chi_S(r^{-1}gr) \text{ if } t = rsr^{-1} \text{ and } r^{-1}gr \in C_G(s), \end{cases}$$

which can be reformulated into the following lemma.

Lemma 8.1.7 *Let us set $\chi_{(s,S)}(a) := \text{tr}(a; X(s, S))$ for $a \in D_kG$. Then*

- $\chi_{(s,S)}(\delta_t g) =$

$$\begin{cases} 0 & \text{if } (t, g) \text{ is not } G - \text{conjugate to } (s, h) \text{ for some } h \in C_G(s), \\ \chi_S(h) & \text{if } (t, g) = r(s, h)r^{-1} \text{ for some } h \in C_G(s) \text{ and some } r \in G. \end{cases}$$

- $\chi^0_{X(s,S)} =$

$$\sum_{\substack{r \in G/C_G(s) \\ h \in C_G(s)}} \chi_S(h^{-1}) r\delta_s hr^{-1} = \sum_{r \in G/C_G(s)} r \left(\sum_{h \in C_G(s)} \chi_S(h^{-1})\delta_s h \right) r^{-1}$$

$$= \sum_{r \in G/C_G(s)} r\delta_s \chi^0_S r^{-1}.$$

Idempotents and characters in characteristic zero.

Let K be a characteristic zero field which is big enough for G. Then, for all $s \in G$, K is a splitting field for $C_G(s)$. It follows from Proposition 8.1.5 that K is a splitting field for $D_K G$, that is, whenever X is an irreducible $D_K G$-module, then $\operatorname{End}_{D_K G}(X) = K\operatorname{Id}_X$.

We admit here that, since $D_K G$ is a split semi-simple algebra, the map

$$\prod_{X \in \operatorname{Irr}(D_K G)} \rho_X : D_K G \longrightarrow \prod_{X \in \operatorname{Irr}(D_K G)} \operatorname{End}_K(X) \tag{8.1.8}$$

is an isomorphism (see for example [Lan65, Chap. XVII]).

Remark 8.1.9   Had we waited until we define a structure of Hopf algebra on $D_K G$ (see below), we could have use methods similar to those used in Chap. 2 to prove (8.1.8). We leave this as an exercise to the reader.

Here (8.1.8) may be written

$$\prod_{\substack{s \in [\operatorname{Cl}(G)] \\ S \in \operatorname{Irr}_K(C_G(s))}} \rho_{X(s,S)} : D_K G \longrightarrow \prod_{\substack{s \in [\operatorname{Cl}(G)] \\ S \in \operatorname{Irr}_K(C_G(s))}} \operatorname{End}_K(X(s,S)),$$

from which it follows that

$$\chi_{D_K G}^{\operatorname{reg}} = \sum_{\substack{s \in [\operatorname{Cl}(G)] \\ S \in \operatorname{Irr}_K(C_G(s))}} \chi_{(s,S)}(1)\chi_{(s,S)},$$

hence, by Lemma 8.1.4,

$$\tau = \sum_{\substack{s \in [\operatorname{Cl}(G)] \\ S \in \operatorname{Irr}_K(C_G(s))}} \frac{\chi_{(s,S)}(1)}{|G|}\chi_{(s,S)} = \sum_{\substack{s \in [\operatorname{Cl}(G)] \\ S \in \operatorname{Irr}_K(C_G(s))}} \frac{\chi_S(1)}{|C_G(s)|}\chi_{(s,S)}. \tag{8.1.10}$$

Let us denote by $E_{(s,S)}$ the central idempotent of $D_K G$ whose image under the above isomorphism is the element with component $\operatorname{Id}_{X(s,S)}$ on the factor associated to (s, S) and 0 on the other factors. Then by definition we have, for all $a \in D_K G$,

$$\chi_{(t,T)}(aE_{(s,S)}) = \begin{cases} \chi_{(s,S)}(a) & \text{if } (t, T) = (s, S), \\ 0 & \text{if not.} \end{cases}$$

By equality (8.1.10) above, we see that for all $a \in D_K G$,

$$\tau(aE_{(s,S)}) = \frac{\chi_S(1)}{|C_G(s)|}\chi_{(s,S)}(a),$$

which, using the isomorphism $\widehat{\tau}$ and Lemma 8.1.7, implies the following proposition.

Proposition 8.1.11

$$E_{(s,S)} = \frac{\chi_S(1)}{|C_G(s)|}\chi^0_{(s,S)} = \frac{\chi_S(1)}{|C_G(s)|}\sum_{r\in G/C_G(s)} r\left(\sum_{h\in C_G(s)}\chi_S(h^{-1})\delta_s h\right)r^{-1}.$$

Definition 8.1.12 For α and β central functions on $D_K G$, we define

$$\langle\alpha,\beta\rangle := \frac{1}{|G|}\sum_{s,g\in G}\alpha(\delta_s g^{-1})\beta(\delta_s g).$$

The equation

$$\chi_{(t,T)}(E_{(s,S)}) = \delta_{(t,T),(s,S)}\chi_{(s,S)}(1)$$

provides a proof of the orthogonality relations expressed in the next corollary.

Corollary 8.1.13 *For $X_{(s,S)}$ and $X_{(t,T)}$ irreducible $D_K G$-modules,*

$$\langle\chi_{(s,S)},\chi_{(t,T)}\rangle = \frac{1}{|G|}\sum_{s,g\in G}\chi_{(s,S)}(\delta_s g^{-1})\chi_{(t,T)}(\delta_s g) = \delta_{(t,T),(s,S)}.$$

Exercise 8.1.14 Check that, if $s = t$, the above orthogonality relations are equivalent to the ordinary orthogonality relations between characters of elements of $\mathrm{Irr}_K(C_G(s))$.

Bases of the space of central functions.

The space $\mathrm{CF}(D_K G, K)$ has two "natural" bases:

- the family of irreducible characters $(\chi_{(s,S)})$,
- the family of characteristic functions γ_C for $C \in \mathrm{Cl}((G \times G)^{\mathrm{com}})$ (see Notation 8.1.1).

Let us compute how to express one basis in terms of the other. For $(s, g) \in (G \times G)^{\mathrm{com}}$, let us denote by $C(s, g)$ its G-conjugacy class.

In the following computation,

- pairs (s, S) are such that s runs over a complete set of representatives for $\mathrm{Cl}(G)$ and $S \in \mathrm{Irr}_K(C_G(s))$,
- while pairs (s, g) are such that s runs over the same complete set of representatives for $\mathrm{Cl}(G)$, and g runs over a complete set of representatives for $\mathrm{Cl}(C_G(s))$.

By Lemma 8.1.7, we know that $\chi_{(s,S)}(\delta_t h) = 0$ unless (t, h) is G-conjugate to (s, g) for some $g \in C_G(s)$, in which case $\chi_{(s,S)}(\delta_t h) = \chi_S(g)$. Hence

$$\chi_{(s,S)} = \sum_{g\in[\mathrm{Cl}(C_G(s))]}\chi_S(g)\gamma_{C(s,g)}. \tag{8.1.15}$$

By the orthogonality relations (see Corollary 8.1.13), we have

$$\gamma_{C(s,g)} = \sum_{(t,T)} \langle \chi_{(t,T)}, \gamma_{C(s,g)} \rangle \chi_{(t,T)} .$$

Since

$$\langle \chi_{(t,T)}, \gamma_{C(s,g)} \rangle = \frac{1}{|G|} \chi_{(t,T)}(\delta_s g^{-1}) |C(s,g)| ,$$

it follows from Lemma 8.1.7 that
$\langle \chi_{(t,T)}, \gamma_{C(s,g)} \rangle =$

$$\begin{cases} 0 & \text{if } (s, g^{-1}) \text{ is not } G - \text{conjugate to } (t, h^{-1}) \text{ for some } h \in C_G(t), \\ \chi_S(g^{-1}) & \text{if } g \in C_G(s) . \end{cases}$$

Since $|C(s,g)| = \dfrac{|G|}{|C_G(s,g)|}$, we deduce from what precedes that

$$|C_G(s,g)| \gamma_{C(s,g)} = \sum_{S \in \mathrm{Irr}_K(C_G(s))} \chi_S(g^{-1}) \chi_{(s,S)} . \qquad (8.1.16)$$

8.2 Hopf Algebras: An Introduction from Scratch

Throughout this section, k is a field.

8.2.1 *Notation for Multiple Tensor Products*

About the flip.

Let A be a k-algebra, finite-dimensional as a k-vector space.

- We have 3 morphisms

$$\iota_{i,j} : A \otimes A \to A \otimes A \otimes A \quad \text{(for } (i, j) = (1, 2), (1, 3), (2, 3))$$

defined by

$$\begin{cases} \iota_{1,2} : & x \otimes y \mapsto x \otimes y \otimes 1 \\ \iota_{1,3} : & x \otimes y \mapsto x \otimes 1 \otimes y \\ \iota_{2,3} : & x \otimes y \mapsto 1 \otimes x \otimes y \end{cases}$$

and for $\xi \in A \otimes A$, we set $\xi_{ij} := \iota_{i,j}(\xi)$.

- We recall that we denote by Φ the automorphism of $A \otimes A$ (the flip, see Sect. 1.3.1) defined by

$$\Phi : x \otimes y \mapsto y \otimes x,$$

and by Id_A (or simply by Id) the identity on A.

We may note that

$$\begin{cases} (\Phi \otimes \mathrm{Id})(x \otimes y \otimes z) = y \otimes x \otimes z, \\ (\mathrm{Id} \otimes \Phi)(x \otimes y \otimes z) = (x \otimes z \otimes y), \end{cases}$$

so

$$(\Phi \otimes \mathrm{Id})(\mathrm{Id} \otimes \Phi)(\Phi \otimes \mathrm{Id})(x \otimes y \otimes z) = (z \otimes y \otimes x),$$

and in particular

$$(\Phi \otimes \mathrm{Id})(\mathrm{Id} \otimes \Phi)(\Phi \otimes \mathrm{Id}) = (\mathrm{Id} \otimes \Phi)(\Phi \otimes \mathrm{Id})(\mathrm{Id} \otimes \Phi).$$

The preceding formulae suggest the following notation (alluding to transpositions):

$$\begin{cases} \underset{12}{\Phi} := (\Phi \otimes \mathrm{Id}) : x \otimes y \otimes z \mapsto y \otimes x \otimes z, \\ \underset{23}{\Phi} := (\mathrm{Id} \otimes \Phi) : x \otimes y \otimes z \mapsto x \otimes z \otimes y, \\ \underset{13}{\Phi} := (\Phi \otimes \mathrm{Id})(\mathrm{Id} \otimes \Phi)(\Phi \otimes \mathrm{Id}) : x \otimes y \otimes z \mapsto z \otimes y \otimes x, \end{cases}$$

and we have the "braid relation"

$$\underset{12}{\Phi} \cdot \underset{23}{\Phi} \cdot \underset{12}{\Phi} = \underset{23}{\Phi} \cdot \underset{12}{\Phi} \cdot \underset{23}{\Phi}.$$

The following formulas are then obvious:

$$\underset{12}{\Phi} \cdot \iota_{1,3} = \iota_{2,3},$$

$$\underset{23}{\Phi} \cdot \iota_{1,2} = \iota_{1,3},$$

$$\underset{13}{\Phi} \cdot \iota_{2,3} = \iota_{1,2}.$$

With several vector spaces.

For X and Y vector spaces, we denote by

$$\Phi_{X,Y} : X \otimes Y \to Y \otimes X \, , \, x \otimes y \mapsto y \otimes x \, ,$$

the flip.

Then it is trivial to check that

$$
\begin{aligned}
\Phi_{X,Y \otimes Z} &= \left(\mathrm{Id}_Y \otimes \Phi_{X,Z}\right) \cdot \left(\Phi_{X,Y} \otimes \mathrm{Id}_Z\right) : X \otimes Y \otimes Z \to Y \otimes Z \otimes X \, , \\
\Phi_{X \otimes Y, Z} &= \left(\Phi_{X,Z} \otimes \mathrm{Id}_Y\right) \cdot \left(\mathrm{Id}_X \otimes \Phi_{Y,Z}\right) : X \otimes Y \otimes Z \to Z \otimes X \otimes Y \, .
\end{aligned}
\tag{8.2.1}
$$

8.2.2 Algebras, Coalgebras, Bialgebras, Hopf Algebras

We concentrate, in a single (long and dense) statement, both a "formal" definition of what is an algebra, then the definition of a coalgebra, which consists in inverting all arrows in the definition of an algebra. Then we give the definition of a bialgebra (which is both an algebra and a coalgebra which are "compatible"), to end with the definition of a special type of bialgebra called Hopf algebra.

As exercises, we let the reader

- define what are morphisms between algebras, between coalgebras, between bialgebras, between Hopf algebras – and define a structure of coalgebra on $A \otimes A$ when A is a coalgebra,
- prove the equivalences between items (i), (ii) and (iii) below in the definition of bialgebras.

Definition 8.2.2 Let A be a finite dimensional k-vector space.

- AN ALGEBRA STRUCTURE on A is the following data (1) and (2).

(1) [MULTIPLICATION] A k-linear morphism

$$
\mu : \begin{cases} A \otimes_k A \to A \, , \\ a_1 \otimes a_2 \mapsto a_1 a_2 \, , \end{cases}
$$

called the multiplication, which is *associative*, that is, the following diagram is commutative:

$$
\begin{array}{ccc}
A \otimes A & \xrightarrow{\ \mu\ } & A \\
{\scriptstyle \mathrm{Id} \otimes \mu} \uparrow & & \uparrow {\scriptstyle \mu} \\
A \otimes A \otimes A & \xrightarrow{\ \mu \otimes \mathrm{Id}\ } & A \otimes A
\end{array}
$$

(2) [UNIT] A k-linear morphism

$$\eta : k \to A$$

called the unit, such that the following diagram is commutative:

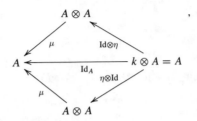

which implies that $\eta(1)$ is the unit element 1_A (simply denoted 1) of A.

(3) [COMMMUTATIVITY] Moreover, the algebra (A, μ, η) is said to be commutative if the following diagram commutes

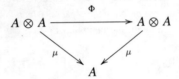

- A COALGEBRA STRUCTURE on A is the following data (1) and (2).

(1) [COMULTIPLICATION] A k-linear morphism

$$\Delta : \begin{cases} A \to A \otimes_k A \,, \\ a \mapsto \sum a_1 \otimes a_2 \,, \end{cases}$$

called the comultiplication, which is *coassociative*, that is, the following diagram is commutative:

$$
\begin{array}{ccc}
A & \xrightarrow{\ \Delta\ } & A \otimes A \\
\Delta \downarrow & & \downarrow \mathrm{Id}\otimes\Delta \\
A \otimes A & \xrightarrow{\ \Delta\otimes\mathrm{Id}\ } & A \otimes A \otimes A
\end{array}
$$

(2) [COUNIT] A k-linear morphism

$$\varepsilon : A \to k \,,$$

called the *counit,* that is, the following diagram is commutative:

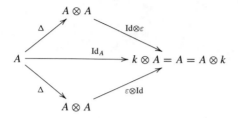

(3) [COCOMMUTATIVITY] Moreover, the coalgebra (A, Δ, ε) is said to be *cocommutative* if the following diagram commutes

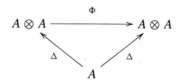

- A BIALGEBRA STUCTURE on A is the data consisting in both an algebra structure (A, μ, η) and a coalgebra structure (A, Δ, ε) which are compatible, that is, they satisfy one of the following equivalent assertions:

(i)   (a) $\eta : k \to A$ is a coalgebra morphism and $\varepsilon : A \to k$ is an algebra morphism,
        (b) the following diagram is commutative:

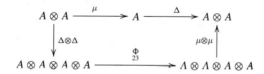

(ii) Δ and ε are algebra morphisms.
(iii) μ and η are coalgebra morphisms.

- A HOPF ALGEBRA STRUCTURE on A is the following data.

(1) A bialgebra structure $(A, \mu, \eta, \Delta, \varepsilon)$ on A.
(2) [ANTIPODE] A k-linear automorphism α of A, called the *antipode*, such that the following diagram is commutative:

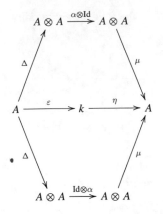

Examples 8.2.3

(1) Let G be a finite group. We define the following five k-linear maps by their action on the basis G of the group algebra kG.

$$\begin{cases} \mu : kG \otimes kG \to kG \ , \ (g_1, g_2) \mapsto g_1 g_2 \, , \\ \eta : k \to kG \ , \ 1 \mapsto 1 \, , \\ \Delta : kG \to kG \otimes kG \ , \ g \mapsto g \otimes g \, , \\ \varepsilon : kG \to k \ , \ g \mapsto 1 \, , \\ \alpha : kG \to kG \ , \ g \mapsto g^{-1} \, . \end{cases}$$

The reader may prove as an exercise that they endow kG with a Hopf algebra structure.

(2) Let V be a finite dimensional k-vector space, and let $T(V)$ be its tensor algebra. We define the following two k-algebra morphisms by their action on any $v \in V$:

$$\begin{cases} \Delta : T(V) \to T(V) \otimes T(V) \ , \ v \mapsto v \otimes 1 + 1 \otimes v \, , \\ \varepsilon : T(V) \to k \ , \ v \mapsto 0 \, , \end{cases}$$

and the antipode of the algebra $T(V)$ by its action on any $v \in V$:

$$\alpha : v \mapsto -v \, .$$

The reader may prove as an exercise that they endow $T(V)$ with a Hopf algebra structure.

⚠ **Attention** ⚠ In the two preceding examples,

- the Hopf algebras are cocommutative,
- their antipode has order 2.

This need not be the case in general, as shown by the Exercise 8.4.14.

The next proposition is stated without proof. The reader may try to prove it as an exercise, or refer to the literature (for example, [Kas95, Part I, Sect. 3]).

Proposition 8.2.4

(1) [DUAL OF A HOPF ALGEBRA] *If $(A, \mu, \eta, \Delta, \varepsilon, \alpha)$ is a Hopf algebra, then applying the dual functor $* = \mathrm{Hom}_k(-, k)$ defines a Hopf algebra $(A^*, \Delta^*, \varepsilon^*, \mu^*, \eta^*, \alpha^*)$.*

(2) [THE ANTIPODE IS AN ANTIAUTOMORPHISM] *Let $(A, \mu, \eta, \Delta, \varepsilon, \alpha)$ be a Hopf algebra. The antipode α is an* antiautomorphism *of the bialgebra A, that is*

(a) • $\alpha \cdot \eta = \eta$,
  • *for all $a_1, a_2 \in A$, $\alpha(a_1 a_2) = \alpha(a_2)\alpha(a_1)$, which can be reformulated as $\alpha \cdot \mu = \mu \cdot \Phi \cdot (\alpha \otimes \alpha)$, and, setting $\mu^{\mathrm{op}} := \mu \cdot \Phi$, it becomes*

$$\alpha \cdot \mu = \mu^{\mathrm{op}} \cdot (\alpha \otimes \alpha).$$

(b) • $\varepsilon \cdot \alpha = \varepsilon$,
  • $\Delta \cdot \alpha = \Phi \cdot (\alpha \otimes \alpha) \cdot \Delta$, *and, setting $\Delta^{\mathrm{op}} := \Phi \cdot \Delta$ it can be rewritten*

$$(\alpha \otimes \alpha) \cdot \Delta = \Delta^{\mathrm{op}} \cdot \alpha.$$

Hopf algebra structure on $D_k G$.

We define on the algebra $D_k G$ a structure of coalgebra, which turns $D_k G$ into a bialgebra, as well as an antipode which turns $D_k G$ into a Hopf algebra.

Definition 8.2.5 The *Drinfeld double of G over k* is the algebra $D_k G$ endowed with the following operations:

(1) [COMULTIPLICATION]

$$\Delta : \begin{cases} D_k G \to D_k G \otimes_k D_k G, \\ \delta_s g \mapsto \displaystyle\sum_{s_1 s_2 = s} \delta_{s_1} g \otimes \delta_{s_2} g, \end{cases}$$

(2) [COUNIT] the linear map

$$\varepsilon : D_k G \to k, \quad \delta_s g \mapsto \delta_{s,1},$$

called the counit for $D_k G$,

(3) [ANTIPODE] the antiautomorphism α of $D_k G$

$$\alpha(\delta_s g) := g^{-1} \delta_{s^{-1}} = \delta_{g^{-1} s^{-1} g} g^{-1},$$

called the antipode.

We leave as an exercise to the reader to check that these operations do turn $D_k G$ into a Hopf algebra.

Remark 8.2.6 Notice (check) that

- the comultiplication Δ on $D_k G$ is *not* cocommutative,
- the antipode α is an involution.

8.2.3 On Algebra Representations of a Hopf Algebra

Recall that, for A a k-algebra, we denote by $_A\mathbf{mod}$ the category of A-modules which are finite dimensional over k. If there is a structure of Hopf algebra on A, the category $_A\mathbf{mod}$ inherits more structure.

Let us first introduce some notation.

- For M a finite dimensional k-vector space, we denote by $M^* := \mathrm{Hom}_k(M, k)$ its dual, and we call *evaluation map* the k-linear map

$$\mathrm{ev}_M : M^* \otimes M \to k \ , \ \varphi \otimes m \mapsto \varphi(m) \,.$$

Through the isomorphism

$$M^* \otimes M \to \mathrm{End}_k(M) \ , \ \varphi \otimes m \mapsto (x \mapsto \varphi(m)x) \,,$$

the map ev becomes the trace tr_M.

- We denote by 1_M the element of $M \otimes M^*$ which corresponds to Id_M through the isomorphism

$$M \otimes M^* \xrightarrow{\ \Phi\ } M^* \otimes M \xrightarrow{\ \sim\ } \mathrm{End}_k(M) \,,$$

(where $\Phi : m \otimes \varphi \mapsto \varphi \otimes m$ is the flip). Then we call *coevaluation map* the k-linear map

$$\mathrm{coev}_M : k \to M \otimes M^* \ , \ 1 \mapsto 1_M \,.$$

Proposition 8.2.7 *Let* $(A, \mu, \eta, \Delta, \varepsilon, \alpha)$ *be a Hopf algebra.*

(1) [TENSOR PRODUCT]

   (a) Given A-modules M and N, then $M \otimes N$ is naturally an $(A \otimes A)$-module. The map

$$(a, m \otimes n) \mapsto \Delta(a)(m \otimes n) \,,$$

   induces on $M \otimes N$ a structure of A-module.

   (b) Given A-modules L, M, N, the canonical morphism

$$(L \otimes M) \otimes N \to L \otimes (M \otimes N) \ , \ (l \otimes m) \otimes n \mapsto l \otimes (m \otimes n)$$

is an isomorphism of A-modules (which then defines the A-module $L \otimes M \otimes N$).

(2) [TRIVIAL MODULE]

(a) The counit $\varepsilon : A \to k$ defines a structure of A-module on k, called the trivial A-module.

(b) The canonical isomorphisms

$$k \otimes M \xrightarrow{\sim} M \xleftarrow{\sim} M \otimes k , \quad \lambda \otimes m \mapsto \lambda m \leftarrow m \otimes \lambda ,$$

are isomorphisms of A-modules (which then identify M, $k \otimes M$, and $M \otimes k$).

(3) [CONTRAGREDIENT MODULE]

(a) The map

$$(a, \varphi) \mapsto (m \mapsto \varphi(\alpha(a)m)) ,$$

endows the dual $M^* := \mathrm{Hom}_k(M, k)$ with a structure of A-module.

(b) The evaluation and coevaluation maps

$$\mathrm{ev} : M^* \otimes M \to k , \quad \varphi \otimes m \mapsto \varphi(m) ,$$
$$\mathrm{coev} : k \to M \otimes M^* , \quad 1 \mapsto 1_M ,$$

are morphisms of A-modules.

(c) If moreover the antipode is an involution, M^{**} is canonically identified with M as an A-module. Then M^* is called the contragredient module.

Sketch of proof

(1)(a) follows from the fact that Δ is an algebra morphism, and (1)(b) is a consequence of the coassociativity.

(2)(a) follows from the fact that ε is an algebra morphism, and (2)(b) is a consequence of the commutativity of the diagram linking counit and comultiplication.

(3)(a) follows from the fact that the antipode α is an antiautomorphism of algebras (see Proposition 8.2.4), and (3)(b) is a consequence of the commutativity of the hexagonal-type diagram defining an antipode.

Finally, (3)(c) is immediate.   □

Remark 8.2.8 In general, for M and N two A-modules, the flip $\Phi_{M,N}$ is not an isomorphism of A-modules. Moreover, the A-modules $M \otimes N$ and $N \otimes M$ need not be isomorphic. The "ribbon structure" defined below will provide such an isomorphism, which moreover will satisfy the analog of (8.2.1).

8.3   The Drinfeld Double as a Ribbon Hopf Algebra

We shall see now that the Drinfeld double of a finite group is endowed with a supplementary structure which provides functorial isomorphisms between $M \otimes N$ and $N \otimes M$ (a substitute for cocommutativity), and which will endow $_{D_kG}\mathbf{mod}$ with the structure of a *ribbon category*.

8.3.1   Universal R-matrix for D_kG

In what follows, Φ denotes the flip

$$\Phi : D_kG \otimes_k D_kG \to D_kG \otimes_k D_kG \ , \ x \otimes y \mapsto y \otimes x \,.$$

The properties described in the following theorem express the fact that R is a *"universal R-matrix"* for the Hopf algebra D_kG (for the notation used in item (3) of the following theorem, the reader may refer to Sect. 8.2.1).

A Hopf algebra with such a universal matrix is said to be *quasitriangular*. As we shall see, this property is a substitute for cocommutativity.

Theorem 8.3.1 *The element $R \in D_kG \otimes_k D_kG$ defined by*

$$R := \sum_{g \in G} \delta_g \otimes g$$

satisfies the following properties (where Id *stands for* Id_{D_kG}*).*

(1) R is invertible. More precisely

$$R^{-1} = (\alpha \otimes \mathrm{Id})(R) \ \ and \ \ R = (\mathrm{Id} \otimes \alpha)(R^{-1}) \,.$$

(2) For all $x \in D_kG$,

$$\Phi(\Delta(x)) = R\Delta(x)R^{-1} \,.$$

(3) $(\Delta \otimes \mathrm{Id})(R) = R_{13}R_{23}$ and $(\mathrm{Id} \otimes \Delta)(R) = R_{13}R_{12} \,.$

Proof

(1) By definition,

$$(\alpha \otimes \mathrm{Id})(R) = \sum_{g \in G}(\delta_{g^{-1}} \otimes g) \,,$$

hence

$$R \cdot (\alpha \otimes \mathrm{Id})(R) = \sum_{g,h \in G} \delta_g \delta_{h^{-1}} \otimes gh = \sum_{g \in G} \delta_g \otimes 1 = 1_{D_k G \otimes D_k G}.$$

(2) For $x = \delta_s g$ (where $s, g \in G$), we have

$$\Phi(\Delta(x)) \cdot R = \Big(\sum_{tu=s} \delta_u g \otimes \delta_t g\Big)\Big(\sum_{h \in G} \delta_h \otimes h\Big) = \sum_{tu=s, h \in G} \delta_u g \delta_h \otimes \delta_t g \delta_h$$

$$= \sum_{h \in G} \delta_{ghg^{-1}} g \otimes \delta_{sgh^{-1}g^{-1}} gh = \sum_{t \in G} \delta_t g \otimes \delta_{st^{-1}} tg,$$

$$R \cdot \Delta(x) = \Big(\sum_{t \in G} \delta_t \otimes t\Big)\Big(\sum_{tu=s} \delta_t g \otimes \delta_u g\Big)$$

$$= \sum_{tu=s, t \in G} \delta_t \delta_t g \otimes t \delta_u g = \sum_{t \in G} \delta_t g \otimes \delta_{st^{-1}} tg,$$

which proves the desired equality.

(3)

$$(\Delta \otimes \mathrm{Id})(R) = (\Delta \otimes \mathrm{Id})\Big(\sum_{g \in G} \delta_g \otimes g\Big) = \sum_{g,s \in G} \delta_s \otimes \delta_{s^{-1}g} \otimes g,$$

while

$$R_{13} R_{23} = \Big(\sum_{s \in G} \delta_s \otimes 1 \otimes s\Big) \times \Big(\sum_{t \in G} 1 \otimes \delta_t \otimes t\Big)$$

$$= \sum_{s,t \in G} \delta_s \otimes \delta_t \otimes st = \sum_{g,s \in G} \delta_s \otimes \delta_{s^{-1}g} \otimes g,$$

proving the desired equality.

The proof of the last equality in (3) is quite analogous. □

Corollary 8.3.2

(1) R satisfies the Yang–Baxter equation, that is

$$R_{12} R_{13} R_{23} = R_{23} R_{13} R_{12}.$$

(2) $(\varepsilon \otimes \mathrm{Id})(R) = (\mathrm{Id} \otimes \varepsilon)(R) = 1_D.$

Proof

(1) By Theorem 8.3.1,

$$R_{12}R_{13}R_{23} = R_{12}(\Delta \otimes \text{Id})(R) \qquad\qquad \text{by (3)}$$
$$= ((\Phi \cdot \Delta) \otimes \text{Id})(R)R_{12} \qquad\qquad \text{by (2)}$$
$$= (\Phi \otimes \text{Id})(R_{13}R_{23})R_{12} = \underset{12}{\Phi}(R_{13}R_{23})R_{12} \quad \text{by (3)}$$
$$= R_{23}R_{13}R_{12}\,.$$

(2)

$$R = (\text{Id} \otimes \text{Id})(R) = (\varepsilon \otimes \text{Id} \otimes \text{Id})(\Delta \otimes \text{Id})(R) \quad \text{(by 8.3.1 (2))}$$
$$= (\varepsilon \otimes \text{Id} \otimes \text{Id})(R_{13}R_{23}) \qquad\qquad \text{(by 8.3.1 (4)(c))}$$
$$= (\varepsilon \otimes \text{Id})(R) \cdot R$$

which implies $(\varepsilon \otimes \text{Id})(R) = 1_D$ since R is invertible. $\qquad\qquad\qquad \square$

Definition 8.3.3 The *Drinfeld element* is the element $u \in D_k G$ defined by

$$u := (\mu \cdot (\alpha \otimes \text{Id}) \cdot \Phi)(R) = \sum_{g \in G} g^{-1}\delta_g = \sum_{g \in G} \delta_g g^{-1}\,.$$

Proposition 8.3.4

(1) u is invertible, and $u^{-1} = \sum_{g \in G} \delta_g g$,
(2) u belongs to the center of the algebra $D_k G$,
(3) $\alpha(u) = u$, and $\varepsilon(u) = 1$,
(4) $\Delta(u) = (u \otimes u)(\Phi(R)R)^{-1} = (\Phi(R)R)^{-1}(u \otimes u)$.

Proof Proofs of the first three assertions are immediate and left to the reader. Let us prove (4). It suffices to check that

$$\Phi(R)R\Delta(u) = u \otimes u\,.$$

Since $\Phi(R) = \sum_{g \in G} g \otimes \delta_g$, we have

$$\Phi(R)R = \sum_{g,h \in G} g\delta_h \otimes \delta_g h\,,$$

and

$$\Phi(R)R\Delta(u) = \sum_{\substack{g,h,s,t,u \in G \\ tu=s}} g\delta_h\delta_t s^{-1} \otimes \delta_g h \delta_u s^{-1}\,.$$

The term $g\delta_h\delta_t s^{-1} \otimes \delta_g h \delta_u s^{-1}$ is zero unless $h = t$ and $huh^{-1} = g$, hence $s = gh$, $t = h, u = h^{-1}gh$, in which case it is equal to $g\delta_h h^{-1}g^{-1} \otimes \delta_g g^{-1} = \delta_{g'}g'^{-1} \otimes \delta_g g^{-1}$ with $g' := ghg^{-1}$, which shows that

$$\Phi(R)R\Delta(u) = \sum_{g,g' \in G} \delta_{g'} g'^{-1} \otimes \delta_g g^{-1} = u \otimes u\,.$$

$\square$

8.3.2 The Category $D_k G\mathbf{mod}$ is a Ribbon Category

For more details about categorical notions introduced here ("braided", "ribbon", "twist", etc.), we refer the reader to [EGNO15].

The category $D_k G\mathbf{mod}$ is a braided category.

Let X and Y be $D_k G$-modules. We define

$$c_{X,Y} : \begin{cases} X \otimes Y \longrightarrow Y \otimes X\,, \\ x \otimes y \mapsto \Phi_{X,Y}(R.(x \otimes y)) = \sum_{g \in G} gy \otimes \delta_g x\,. \end{cases}$$

It is clear that $c_{X,Y}$ is invertible and more precisely

$$c_{X,Y}^{-1} = R^{-1} \Phi_{Y,X} : y \otimes x \mapsto \sum_{g \in G} \delta_{g^{-1}} x \otimes gy\,.$$

Properties stated in the following Proposition (compare to (8.2.1)) amount to saying (see for example [EGNO15, Chap. 8]) that the category $D_k G\mathbf{mod}$ is a *braided category*.

Proposition 8.3.5

(1) $c_{X,Y}$ is an isomorphism of $D_k G$-modules.
(2) The family $(c_{X,Y})$ satisfies the conditions

$$\begin{cases} c_{X,Y\otimes Z} = (\mathrm{Id}_Y \otimes c_{X,Z}) \cdot (c_{X,Y} \otimes \mathrm{Id}_Z) \\ \qquad = X \otimes Y \otimes Z \to Y \otimes X \otimes Z \to Y \otimes Z \otimes X\,, \\ c_{X\otimes Y,Z} = (c_{X,Z} \otimes \mathrm{Id}_Y) \cdot (\mathrm{Id}_X \otimes c_{Y,Z}) \\ \qquad = X \otimes Y \otimes Z \to X \otimes Z \otimes Y \to Z \otimes X \otimes Y\,. \end{cases}$$

Proof It is left as an exercise to the reader, who may either use the properties proved above about R, or refer to the "more concrete" description of the morphisms $c_{X,Y}$ below. $\square$

The category $D_k G\mathbf{mod}$ is a ribbon category.

For any $D_k G$-module X, we define the automorphism of $D_k G$-modules

$$\theta_X : X \longrightarrow X , \; x \mapsto u^{-1}x .$$

The next proposition expresses the fact that $D_k G$ is a *ribbon Hopf algebra* (see [EGNO15, Sect. 8.11]), which implies that $_{D_k G}\mathbf{mod}$ is a *ribbon category* (see [EGNO15, Sect. 8.10]).

Proposition 8.3.6 *For all $D_k G$-modules X and Y, we have*

(1) $\theta_{X \otimes Y} = (\theta_X \otimes \theta_Y) \cdot c_{Y,X} \cdot c_{X,Y} .$
(2) $(\theta_X)^* = \theta_{X^*} .$

Proof

(1) follows from the equality $\Delta(u) = (\Phi(R)R)^{-1}(u \otimes u)$ (see item 4 of Proposition 8.3.4).
(2) Follows from the fact that $\alpha(u) = u$ (see item 3 of Proposition 8.3.4). $\qquad\square$

8.3.3 A Description of $_{D_k G}\mathbf{mod}$ as a Ribbon Category

TENSOR PRODUCT

For $X = \bigoplus_{s \in G} {}_s X$ and $Y = \bigoplus_{s \in G} {}_s Y$ two G-graded kG-modules, we have

$$_s(X \otimes Y) := \sum_{tu=s} {}_t X \otimes {}_u Y$$

and for $x \otimes y \in {}_t X \otimes {}_u Y$ and $g \in G$,

$$g(x \otimes y) := gx \otimes gy ,$$

an element of $_{gt} X \otimes {}_{gu} Y \subset {}_{gs}(X \otimes Y) .$

DUAL

- For X a G-graded kG-module, we have

$$X^* = \bigoplus_{s \in G} {}_s(X^*) \quad \text{where } {}_s(X^*) = ({}_{s^{-1}}X)^* .$$

- Note that

$$_s(X \otimes X^*) = \bigoplus_{tu=s} {}_t X \otimes {}_u(X^*) = \bigoplus_{tu=s} {}_t X \otimes ({}_{u^{-1}}X)^* ,$$

and in particular

$$_1(X \otimes X^*) = \bigoplus_{s \in G} {}_s X \otimes {}_{s^{-1}}(X^*) = \bigoplus_{s \in G} {}_s X \otimes ({}_s X^*) .$$

Thus ev_X and coev_X are defined (with obvious notation) by

$$
\begin{cases}
\mathrm{ev}_X : X^* \otimes X \to k : \begin{cases} {}_s(X \otimes X^*) \to 0 \ \ \text{if } s \neq 1, \\[2mm] {}_1(X \otimes X^*) \to k : \begin{cases} {}_sX \otimes {}_s(X^*) \to k, \\[1mm] x \otimes \varphi \mapsto \langle x, \varphi \rangle. \end{cases} \end{cases} \\[10mm]
\mathrm{coev}_X : k \to X \otimes X^* : \begin{cases} k \mapsto \bigoplus_{s \in G} {}_sX \otimes {}_s(X^*), \\[3mm] 1 \mapsto \bigoplus_{s \in G} \mathrm{Id}_{{}_sX}. \end{cases}
\end{cases}
$$

RIBBON CATEGORY STRUCTURE

- The morphism

$$
c_{X,Y} : X \otimes Y = \bigoplus_{s \in G} \bigoplus_{tu=s} {}_tX \otimes {}_uY \longrightarrow Y \otimes X
$$

is such that

$$
c_{X,Y} : \begin{cases} {}_s(X \otimes Y) \longrightarrow {}_s(Y \otimes X), \ \ \text{with} \\[2mm] {}_tX \otimes {}_uY \longrightarrow {}_{tut^{-1}}Y \otimes {}_tX, \\[2mm] x \otimes y \mapsto ty \otimes x. \end{cases}
$$

- Finally,

$$
\theta_X : \begin{cases} X \longrightarrow X, \ \ {}_sX \longrightarrow {}_sX, \\[2mm] x \mapsto sx. \end{cases}
$$

Exercise 8.3.7 Using the above notation, prove that θ is a twist, that is, for all $D_k G$-modules X and Y, we have

$$
\theta_{X \otimes Y} = \theta_X \circ \theta_Y \circ c_{Y,X} \circ c_{X,Y}.
$$

The graded characters.

For $X = \bigoplus_{s \in G} {}_sX$ a $D_k G$-module, and for $g \in G$, the contribution of ${}_sX$ to the trace of g on X is zero unless $s \in C_G(g)$, and in that case ${}_sX$ is stable under g. Hence

$$
\mathrm{tr}(g; X) = \sum_{s \in C_G(g)} \mathrm{tr}(g; {}_sX). \tag{8.3.8}
$$

Definition 8.3.9 For $X = \bigoplus_{s \in G}\, {}_sX$ a D_kG-module, we call G-graded character of X the function

$$\mathrm{grchar}_G(X) : G \to kG \ , \quad g \mapsto \mathrm{grchar}_G(X)(g) = \sum_{s \in C_G(g)} \mathrm{tr}(g; \,{}_sX)s \,.$$

Note that, for all $g \in G$, $\mathrm{grchar}_G(X)(g) \in ZkC_G(g)$.

Indeed, it suffices to check that for all $h \in C_G(g)$, we have

$$\mathrm{tr}(g; \,{}_sX) = \mathrm{tr}(g; \,{}_{hsh^{-1}}X) \,.$$

But ${}_{hsh^{-1}}X = {}^h{}_sX$, hence

$$\mathrm{tr}(g; \,{}_{hsh^{-1}}X) = \mathrm{tr}(h^{-1}gh; \,{}_sX) = \mathrm{tr}(g; \,{}_sX) \,,$$

since $h \in C_G(g)$.

Thus the value at $t \in G$ of the graded character of $X(s, S)$ is the element of $ZkC_G(t)$ (see Definition 8.3.9) defined by

$$\mathrm{grchar}(X(s, S))(t) = \sum_{u \in C_G(t) \cap \mathrm{Cl}(s)} \mathrm{tr}(t; \,{}_uX(s, S))u$$

$$= \sum_{\substack{r \in [G/C_G(s)] \\ rsr^{-1} \in C_G(t)}} \chi_S(t)^r s = \sum_{\substack{r \in [G/C_G(s)] \\ r^{-1}tr \in C_G(s)}} \chi_S(r^{-1}tr)^r s \,.$$

Now, if $X = \bigoplus_{t \in G}\, {}_tX$ and $Y = \bigoplus_{u \in G}\, {}_uY$ are two D_kG-modules,

$$\mathrm{tr}(g; X \otimes Y) = \sum_{t,u \in G} \mathrm{tr}(g; \,{}_tX \otimes\, {}_uY)$$

$$= \sum_{t,u \in G} \mathrm{tr}(g; \,{}_tX)\mathrm{tr}(g; \,{}_uY) \tag{8.3.10}$$

$$= \sum_{t,u \in C_G(g)} \mathrm{tr}(g; \,{}_tX)\mathrm{tr}(g; \,{}_uY) \,.$$

It follows from formula (8.3.10) that:

Lemma 8.3.11 *Whenever X and Y are D_kG-modules,*

$$\mathrm{grchar}_G(X \otimes Y) = \mathrm{grchar}_G(X)\mathrm{grchar}_G(Y) \,.$$

Characters of the Grothendieck Ring.

Until the end of this chapter, K is a characteristic zero (commutative) field which is supposed to be big enough for G.

Choose an irreducible $KC_G(t)$-module T, and apply the corresponding central character ω_T to $\mathrm{grchar}(X(s, S))(t)$. We get

$$\omega_T(\mathrm{grchar}(X(s, S))(t)) = \frac{1}{\chi_T(1)} \sum_{\substack{r \in [G/C_G(s)] \\ r^{-1}tr \in C_G(s)}} \chi_S(r^{-1}tr)\chi_T(rsr^{-1}). \qquad (8.3.12)$$

Let us introduce the following notation:

$$G(s, t) := \{\sigma \in G \mid \sigma s \sigma^{-1} t = t \sigma s \sigma^{-1}\}.$$

Notice that $\sigma \in G(s, t)$ if and only if $\sigma^{-1} \in G(t, s)$, and also that $C_G(s)$ acts by right multiplication on $G(s, t)$ while $C_G(t)$ acts on it by left multiplication.

The formula (8.3.12) above may be rewritten

$$\omega_T(\mathrm{grchar}(X(s, S))(t)) =$$
$$\frac{1}{\chi_T(1)}\frac{1}{|C_G(s)|} \sum_{\sigma \in G(s,t)} \chi_S(\sigma^{-1}t\sigma)\chi_T(\sigma s \sigma^{-1}) =$$
$$\frac{1}{\chi_{X_{(t,T)}}(1)}\frac{|G|}{|C_G(s)||C_G(t)|} \sum_{\sigma \in G(s,t)} \chi_S(\sigma^{-1}t\sigma)\chi_T(\sigma s \sigma^{-1}). \qquad (8.3.13)$$

Let us denote by $\mathrm{Gr}(D_K G)$ the Grothendieck ring of $_{D_K G}\mathbf{mod}$. This is a free $\mathbb{Z}$-module, with basis $(X)_{X \in \mathrm{Irr}(D_K G)}$. Since $_{D_K G}\mathbf{mod}$ is also endowed with a tensor product, $\mathrm{Gr}(D_K G)$ is a *ring*.

The following proposition is then an immediate consequence of what precedes, and of Lemma 8.3.11.

Proposition 8.3.14 *For a given irreducible $D_K G$-module $X_{(t,T)}$, the formula*

$$X_{(s,S)} \mapsto \omega_T(\mathrm{grchar}(X(s, S))(t)) =$$
$$\frac{1}{\chi_{X_{(t,T)}}(1)}\frac{|G|}{|C_G(s)||C_G(t)|} \sum_{\sigma \in G(s,t)} \chi_S(\sigma^{-1}t\sigma)\chi_T(\sigma s \sigma^{-1})$$

defines a ring morphism $\mathrm{Gr}(D_K G) \to K$.

The twist.

Let us compute the twist $\theta_{(s,S)} := \theta_{X(s,S)}$ acting on an irreducible module.

Lemma 8.3.15 *For $s \in G$ and $S \in \mathrm{Irr}_K(C_G(s))$, the twist $\theta_{(s,S)}$ is the multiplication by s.*

In particular if $\mathrm{End}_{KC_G(s)}(S) = K\mathrm{Id}_S$, then $\theta_{(s,S)}$ is the scalar multiplication by $\chi_S(s)/\chi_S(1)$.

The numbers $s_{X,Y}$.

Let X and Y be $D_K G$-modules. For $x \in {}_t X$ and $y \in {}_u Y$, we have $c_{X,Y} : x \otimes y \mapsto ty \otimes x$, hence

$$c_{Y,X} c_{X,Y} : x \otimes y \mapsto tut^{-1} x \otimes ty \in {}_{tutu^{-1}t^{-1}} X \otimes {}_{tut^{-1}} Y,$$

and ${}_t X \otimes {}_u Y$ contributes to the trace of $c_{Y,X} c_{X,Y}$ if and only if

$$_{tutu^{-1}t^{-1}} X = {}_t X \quad \text{and} \quad _{tut^{-1}} Y = {}_u Y,$$

that is, $tu = ut$. In which case

$$c_{Y,X} c_{X,Y} : x \otimes y \mapsto ux \otimes ty.$$

The next lemma is then immediate.

Lemma 8.3.16

$$\operatorname{tr}(c_{Y,X} c_{X,Y}; X \otimes Y) = \sum_{\substack{u,t \in G \\ ut = tu}} \chi_{t X}(u) \chi_{u Y}(t).$$

Notation 8.3.17 We set

$$s_{X,Y} := \operatorname{tr}(c_{Y,X} c_{X,Y}; X \otimes Y).$$

Exercise 8.3.18 Prove that

(1) the matrix $(s_{X,Y})_{X,Y \in \operatorname{Irr}(D_K G)}$ is symmetric,
(2) $s_{X,1} = s_{1,X} = \dim X$,
(3) $s_{X^*,Y^*} = s_{X,Y}$.

Assume that X and Y are irreducible $D_K G$-modules, and more precisely that

$$X = \operatorname{Ind}_{C_G(g)}^G S \quad \text{and} \quad Y = \operatorname{Ind}_{C_G(h)}^G T,$$

where S is an irreducible $K C_G(g)$-module and T is an irreducible $K C_G(h)$-module. We set

$$s_{X,Y} := s_{(g,S),(h,T)}.$$

Then

$$_t X \neq 0 \Leftrightarrow t = rgr^{-1} \text{ for some } r \in G,$$

$$\text{thus } {}_t X = {}^r S \text{ and } \chi_{t X}(a) = \chi_S(r^{-1} ar) \text{ for } a \in G,$$

$$_u Y \neq 0 \Leftrightarrow u = shs^{-1} \text{ for some } s \in G,$$

$$\text{thus } _u Y = {}^s T \text{ and } \chi_{_u Y}(b) = \chi_T(s^{-1}bs) \text{ for } b \in G.$$

Lemma 8.3.16 implies

$$\mathbf{s}_{(g,S),(h,T)} = \sum_{\substack{r \in [G/C_G(g)], s \in [G/C_G(h)] \\ [shs^{-1}, rgr^{-1}]=1}} \chi_S(r^{-1}shs^{-1}r)\chi_T(s^{-1}rhr^{-1}s).$$

In the above summation, we may replace $r^{-1}s$ by $\alpha^{-1}(r^{-1}s)\beta$, where $\alpha \in C_G(g)$ and $\beta \in C_G(h)$, without changing the value of any element of the sum. Summing over α and β, we then get

$$\mathbf{s}_{(g,S),(h,T)} = \frac{1}{|C_G(g)||C_G(h)|} \sum_{\substack{r,s \in G \\ [shs^{-1}, rgr^{-1}]=1}} \chi_S(r^{-1}shs^{-1}r)\chi_T(s^{-1}rgr^{-1}s).$$

We recall that

$$G(g, h) := \{\sigma \in G \mid g\sigma h\sigma^{-1} = \sigma h\sigma^{-1}g\}.$$

Then $r^{-1}s \in G(g, h)$, and the above formula provides the following proposition.

Proposition 8.3.19

$$\mathbf{s}_{(g,S),(h,T)} = \frac{|G|}{|C_G(g)||C_G(h)|} \sum_{\sigma \in G(g,h)} \chi_S(\sigma h\sigma^{-1})\chi_T(\sigma^{-1}g\sigma).$$

Comparing the previous proposition with Proposition 8.3.14, we get the following statement.

Corollary 8.3.20 *For each irreducible $D_K G$-module Y, the linear map $\mathbf{h}_Y$ defined on an irreducible $D_K G$-module X by*

$$\mathbf{h}_Y : X \mapsto \frac{1}{\chi_Y(1)} \mathbf{s}_{X,Y},$$

induces a ring morphism $\mathrm{Gr}(D_K G) \to K$.

Remark 8.3.21 For another proof of the preceding corollary, along the more general lines of [EGNO15, Sect. 8.1], see Exercise 8.4.16.

8.4 Action of $GL_2(\mathbb{Z}/e_G\mathbb{Z})$

From now on we denote by K a characteristic zero (commutative) field which is big enough for G.

We recall that e_G denotes the lcm of the orders of the elements of G.

8.4.1 The Automorphisms $\mathbf{S}$, $\mathbf{\Omega}$, $\mathbf{\Delta}_n$

We shall define some automorphisms of the K-vector space $\mathrm{CF}(D_K G, K)$ (denoted shortly $\mathrm{CF}(D_K G)$) of central functions from $D_K G$ to K.

We shall describe those automorphisms by their actions on two bases of $\mathrm{CF}(D_K G)$:

- the basis $(\chi_{(s,S)})$ of irreducible characters, where s runs over a complete set of representatives of conjugacy classes of G and $S \in \mathrm{Irr}_K(C_S(s))$,
- the basis $(|C_G(s,t)|\gamma_{C(s,t)})$, where (s,t) runs over a complete set of representatives of G-conjugacy classes of commuting pairs in G (and $C(s,t)$ denotes the conjugacy class of (s,t)).

Definition 8.4.1 *(Remarkable endomorphisms of $\mathrm{CF}(D_K G)$)*

(1) Let $\mathbf{S}$ be the endomorphism of $\mathrm{CF}(D_K G)$ defined by the matrix $(\frac{\mathbf{S}_{X,Y}}{|G|})_{X,Y\in\mathrm{Irr}(D_K G)}$, that is

$$\mathbf{S} : \chi_X \mapsto \sum_{Y\in\mathrm{Irr}(D_K G)} \frac{\mathbf{S}_{X,Y}}{|G|}\chi_Y ,$$

or, in other words,

$$\mathbf{S} : \chi_{(s,S)} \mapsto \sum_{(t,T)} \frac{\mathbf{S}_{(s,S),(t,T)}}{|G|}\chi_{(t,T)}$$

(where s runs over a complete set of representatives of G-conjugacy classes and $S \in \mathrm{Irr}_K(C_G(s))$).

(2) Let $\mathbf{\Omega}$ be the endomorphism of $\mathrm{CF}(D_K G)$ defined by

$$\mathbf{\Omega} : \chi_X \mapsto \theta_X \cdot \chi_X ,$$

or, in other words,

$$\mathbf{\Omega} : \chi_{(s,S)} \mapsto \frac{\chi_S(s)}{\chi_S(1)}\chi_{(s,S)}$$

(where s runs over a complete set of representatives of G-conjugacy classes and $S \in \mathrm{Irr}_K(C_G(s))$).

(3) For $n \in (\mathbb{Z}/e_G\mathbb{Z})^\times$, let $\mathbf{\Delta}_n$ be the endomorphism of $\mathrm{CF}(D_K G)$ defined by (see 3.5.20)

$$\mathbf{\Delta}_n : \chi_X \mapsto \chi_{^nX},$$

or, in other words,

$$\mathbf{\Delta}_n : \chi_{(s,S)} \mapsto \chi_{(s,^nS)}.$$

The following proposition is inspired by [DM85, Chap. VII, Sect. 3].

François Digne (born 1949) & Jean Michel (born 1951)

Proposition 8.4.2 *Whenever (s, t) is a commuting pair of elements of G,*

(1) $\mathbf{S} : |C_G(s, t)|\gamma_{C(s,t)} \mapsto |C_G(t, s^{-1})|\gamma_{C(t,s^{-1})}$.
(2) $\mathbf{\Omega} : |C_G(s, t)|\gamma_{C(s,t)} \mapsto |C_G(s, ts^{-1})|\gamma_{C(s,ts^{-1})}$.
(3) $\mathbf{\Delta}_n : |C_G(s, t)|\gamma_{C(s,t)} \mapsto |C_G(s, t^n)|\gamma_{C(s,t^n)}$ *for $n \in (\mathbb{Z}/e_G\mathbb{Z})^\times$.*

Proof
(1) We have

$$\mathbf{S}(|C_G(s, t)|\gamma_{C(s,t)}) = \sum_{S \in \mathrm{Irr}_K(C_G(s))} \chi_S(t^{-1})\mathbf{S}(\chi_{(s,S)})$$

$$= \sum_{S \in \mathrm{Irr}_K(C_G(s))} \chi_S(t^{-1}) \sum_{(u,U)} \frac{\mathbf{S}_{(s,S),(u,U)}}{|G|}\chi_{(u,U)},$$

hence

$$\mathbf{S}(|C_G(s, t)|\gamma_{C(s,t)}) =$$

$$\sum_{S \in \mathrm{Irr}_K(C_G(s))} \chi_S(t^{-1}) \sum_{(u,U)} \frac{1}{|C_G(s)||C_G(u)|} \sum_{\sigma \in G(s,u)} \chi_S(\sigma u \sigma^{-1})\chi_U(\sigma^{-1}s\sigma)\chi_{(u,U)}.$$

We set $(v, V) := \sigma(u, U)\sigma^{-1}$, and change

$$\frac{1}{|C_G(u)|} \sum_{(u,U)} \sum_{\sigma \in G(s,u)} \quad \text{into} \quad \sum_{(v,V)} \cdot$$

We get:

$$S(|C(s,t)|\gamma_{C(s,t)})$$

$$= \sum_{S \in \mathrm{Irr}_K(C_G(s))} \chi_S(t^{-1}) \frac{1}{|C_G(s)|} \sum_{\substack{v \in C_G(s) \\ V \in \mathrm{Irr}_K(C_G(v))}} \chi_V(s)\chi_S(v)\chi_{(v,V)}$$

$$\sum_{\substack{v \in C_G(s) \\ V \in \mathrm{Irr}_K(C_G(v))}} \chi_V(s) \left(\frac{1}{|C_G(s)|} \sum_{S \in \mathrm{Irr}_K(C_G(s))} \chi_S(t^{-1})\chi_S(v) \right) \chi_{(v,V)} \cdot$$

By the usual orthogonality relations,

$$\frac{1}{|C_G(s)|} \sum_{S \in \mathrm{Irr}_K(C_G(s))} \chi_S(v)\chi_S(t^{-1})$$

$$= \begin{cases} \dfrac{|C_G(v,s)|}{|C_G(s)|} & \text{if } v \text{ and } t \text{ are conjugate in } C_G(s), \\ 0 & \text{if not.} \end{cases}$$

If v and t are conjugate in $C_G(s)$, we may replace

$$\sum_{\substack{v \in C_G(s) \\ V \in \mathrm{Irr}_K(C_G(v))}} \frac{|C_G(v,s)|}{|C_G(s)|} \quad \text{by} \quad \sum_{T \in \mathrm{Irr}_K(C_G(t))} 1,$$

and we get

$$S(|C_G(s,t)|\gamma_{C(s,t)}) = \sum_{T \in \mathrm{Irr}_K(C_G(t))} \chi_T(s)\chi_{(t,T)},$$

hence

$$S(|C_G(s,t)|\gamma_{C(s,t)}) = |C_G(t,s^{-1})|\gamma_{C(t,s^{-1})} \cdot$$

(2) Indeed,

$$\Omega(|C_G(s,t)|\gamma_{C(s,t)}) = \sum_{S \in \mathrm{Irr}_K(C_G(s))} \chi_S(t^{-1})\Omega(\chi_{(s,S)})$$

$$= \sum_{S \in \mathrm{Irr}_K(C_G(s))} \chi_S(t^{-1})\frac{\chi_S(s)}{\chi_S(1)}\chi_{(s,S)} = \sum_{S \in \mathrm{Irr}_K(C_G(s))} \chi_S(t^{-1}s)\chi_{(s,S)}$$

$$= |C_G(s,ts^{-1})|\gamma_{C(s,ts^{-1})} \cdot$$

(3) We have

$$
\begin{aligned}
\Delta_n(|C_G(s,t)|\gamma_{C(s,t)}) &= \sum_{S\in\mathrm{Irr}_K(C_G(s))} \chi_S(t^{-1})\Delta_n(\chi_{(s,S)}) \\
&= \sum_{S\in\mathrm{Irr}_K(C_G(s))} \chi_S(t^{-1})\chi_{(s,{}^nS)} = \sum_{S\in\mathrm{Irr}_K(C_G(s))} \chi_S(t^{-n})\chi_{(s,S)} \\
&= |C_G(s,t^n)|\gamma_{C(s,t^n)}\,.
\end{aligned}
$$

$\square$

Corollary 8.4.3 *For all (s, S), we have*

$$
\mathbf{S}^2 : \chi_{(s,S)} \mapsto \chi^*_{(s,S)} = \chi_{(s^{-1},S^*)}\,.
$$

Proof Since (see (8.1.15))

$$
\chi_{(s,S)} = \sum_{g\in[\mathrm{Cl}(C_G(s))]} \chi_S(g)\gamma_{C(s,g)}\,,
$$

we have

$$
\begin{aligned}
\mathbf{S}^2(\chi_{(s,S)}) &= \sum_{g\in[\mathrm{Cl}(C_G(s))]} \chi_S(g)\mathbf{S}^2(\gamma_{C(s,g)}) = \sum_{g\in[\mathrm{Cl}(C_G(s))]} \chi_S(g)\gamma_{C(s^{-1},g^{-1})} \\
&= \sum_{h\in[\mathrm{Cl}(C_G(s^{-1}))]} \chi_{S^*}(h)\gamma_{C(s^{-1},h)} = \chi_{(s^{-1},S^*)}\,.
\end{aligned}
$$

$\square$

8.4.2   Action of $GL_2(\mathbb{Z}/e_G\mathbb{Z})$

Definition 8.4.4 The group $GL_2(\mathbb{Z}/e_G\mathbb{Z})$ acts naturally on the set $(G \times G)^{\mathrm{com}}$ as follows. For $\begin{pmatrix} a & b \\ c & d \end{pmatrix} \in GL_2(\mathbb{Z}/e_G\mathbb{Z})$ and $(s, t) \in (G \times G)^{\mathrm{com}}$, we set

$$
\begin{pmatrix} a & b \\ c & d \end{pmatrix}.(s, t) := (s^a t^b, s^c t^d)\,.
$$

The next property is trivial.

Lemma 8.4.5 *The action of $GL_2(\mathbb{Z}/e_G\mathbb{Z})$ on $(G \times G)^{\mathrm{com}}$ commutes with the conjugation action by G, hence induces an action on the set of classes $\mathrm{Cl}((G \times G)^{\mathrm{com}})$, which preserves the cardinality of classes.*

Thus $GL_2(\mathbb{Z}/e_G\mathbb{Z})$ acts on $CF(D_KG)$ by permuting the elements of the basis $(|C_G(s,t)|\gamma_{C(s,t)})$, and in particular

$$\begin{pmatrix} 0 & 1 \\ -1 & 0 \end{pmatrix} : |C_G(s,t)|\gamma_{C(s,t)} \mapsto |C_G(t,s^{-1})|\gamma_{C(t,s^{-1})},$$

$$\begin{pmatrix} 1 & 0 \\ -1 & 1 \end{pmatrix} : |C_G(s,t)|\gamma_{C(s,t)} \mapsto |C_G(s,ts^{-1})|\gamma_{C(s,ts^{-1})},$$

$$\begin{pmatrix} 1 & 0 \\ 0 & n \end{pmatrix} : |C_G(s,t)|\gamma_{C(s,t)} \mapsto |C_G(s,t^n)|\gamma_{C(s,t^n)}.$$

The matrices $\begin{pmatrix} 0 & 1 \\ -1 & 0 \end{pmatrix}, \begin{pmatrix} 1 & 0 \\ -1 & 1 \end{pmatrix}, \begin{pmatrix} 1 & 0 \\ 0 & n \end{pmatrix}$ $(n \in (\mathbb{Z}/e_G\mathbb{Z})^\times)$ generate $GL_2(\mathbb{Z}/e_G\mathbb{Z})$. Hence the next proposition follows from Proposition 8.4.2.

Proposition 8.4.6 *The map*

$$\begin{pmatrix} 0 & 1 \\ -1 & 0 \end{pmatrix} \mapsto \mathbf{S}, \quad \begin{pmatrix} 1 & 0 \\ -1 & 1 \end{pmatrix} \mapsto \mathbf{\Omega}, \quad \begin{pmatrix} 1 & 0 \\ 0 & n \end{pmatrix} \mapsto \mathbf{\Delta}_n,$$

defines an operation of $GL_2(\mathbb{Z}/e_G\mathbb{Z})$ *on* $CF(D_KG)$.

We set

$$\mathbf{Sh} := \mathbf{S\Omega S}^{-1},$$

which corresponds to the matrix $\begin{pmatrix} 1 & 1 \\ 0 & 1 \end{pmatrix} \in GL_2(\mathbb{Z}/e_G\mathbb{Z})$. Then (because the corresponding relations hold in $GL_2(\mathbb{Z})$) we have the following set of relations.

Proposition 8.4.7

(1) $\mathbf{\Omega Sh\Omega} = \mathbf{Sh\Omega Sh}$,
(2) $\mathbf{S}^2 = (\mathbf{Sh} \cdot \mathbf{\Omega})^3 = (\mathbf{\Omega} \cdot \mathbf{Sh})^3$,
(3) $[\mathbf{S}^2, \mathbf{\Omega}] = [\mathbf{S}^2, \mathbf{Sh}] = 1$,
(4) $\mathbf{\Delta} \cdot \mathbf{S} \cdot \mathbf{\Delta} = \mathbf{S}^{-1}$.

8.4.3 The Verlinde Formula

Let us first notice this reformulation of Corollary 8.4.3.

Lemma 8.4.8 *For* $X, Y \in \mathrm{Irr}(D_KG)$,

$$\sum_{Z \in \mathrm{Irr}(D_KG)} \mathbf{s}_{X,Z}\mathbf{s}_{Y,Z} = |G|\delta_{X,Y^*}.$$

Notation 8.4.9 For $Y, Z, W \in \mathrm{Irr}(D_K G)$, we define the integers $N_{Y,Z}^W$ by

$$Y \otimes Z \simeq \bigoplus_{W \in \mathrm{Irr}(D_K G)} W^{N_{Y,Z}^W},$$

or, using the representatives $[X]$ of the irreducible modules X in the Grothendieck ring $\mathrm{Gr}(D_K G)$,

$$[Y \otimes Z] = \sum_{W \in \mathrm{Irr}(D_K G)} N_{Y,Z}^W [W].$$

Applying Corollary 8.3.20, since

$$\mathbf{h}_X(Y \otimes Z) = \mathbf{h}_X(Y)\mathbf{h}_X(Z),$$

we get

$$\frac{\mathbf{s}_{X,Y}\mathbf{s}_{X,Z}}{\dim X} = \sum_{W \in \mathrm{Irr}(D_K G)} N_{Y,Z}^W \mathbf{s}_{X,W}. \tag{8.4.10}$$

Proposition 8.4.11 *(Verlinde Formula) For $Y, Z, W \in \mathrm{Irr}(D_K G)$,*

$$|G|^2 N_{Y,Z}^W = \sum_{X \in \mathrm{Irr}(D_K G)} \frac{\mathbf{s}_{X,Y}\mathbf{s}_{X,Z}\mathbf{s}_{X,W^*}}{\dim X}.$$

Proof Multiplying both sides of Eq. 8.4.10 above by $\mathbf{s}_{X,W^*}$ gives

$$\frac{\mathbf{s}_{X,Y}\mathbf{s}_{X,Z}\mathbf{s}_{X,W^*}}{\dim X} = \sum_{W \in \mathrm{Irr}(D_K G)} N_{Y,Z}^W \mathbf{s}_{X,W}\mathbf{s}_{X,W^*}.$$

Now summing over X,

$$\sum_{X \in \mathrm{Irr}(D_K G)} \frac{\mathbf{s}_{X,Y}\mathbf{s}_{X,Z}\mathbf{s}_{X,W^*}}{\dim X} = \sum_{W \in \mathrm{Irr}(D_K G)} N_{Y,Z}^W \sum_{X \in \mathrm{Irr}(D_K G)} \mathbf{s}_{X,W}\mathbf{s}_{X,W^*},$$

and by Lemma 8.4.8, we get

$$\sum_{X \in \mathrm{Irr}(D_K G)} \frac{\mathbf{s}_{X,Y}\mathbf{s}_{X,Z}\mathbf{s}_{X,W^*}}{\dim X} = |G|^2 N_{Y,Z}^W.$$

$\square$

Remark 8.4.12 Let us set $\mathbf{S}_{X,Y} := \dfrac{\mathbf{s}_{X,Y}}{|G|}$. Since Corollary 8.4.3 may be reformulated

$$\mathbf{S}_{X,Y}^{-1} = \mathbf{S}_{X,Y^*},$$

and since dim $X = \mathbf{s}_{X,1}$, Verlinde formula (8.4.11) may as well be written

$$N^W_{Y,Z} = \sum_{X \in \mathrm{Irr}(D_K G)} \frac{\mathbf{S}_{X,Y}\mathbf{S}_{X,Z}\mathbf{S}^{-1}_{X,W}}{\mathbf{S}_{X,1}}. \tag{8.4.13}$$

In particular, the expression on the right hand side of the above formula is a *non-negative integer* for all $Y, Z, W \in \mathrm{Irr}(D_K G)$.

More Exercises on Chap. 8

Exercise 8.4.14 Let k be a field of characteristic not 2. Consider the 4-dimensional k-vector space A with basis $(1, n, s, n')$.

(1) Prove that the rules

$$n^2 = 0, \ s^2 = 1, \ sn = n' = -ns,$$

define a k-algebra structure on A.

(2) Prove that the algebra morphisms Δ and ε defined by

$$\Delta : \begin{cases} A \to A \otimes A, \\ n \mapsto n \otimes 1 + s \otimes n, \\ s \mapsto s \otimes s, \end{cases} \quad \text{and} \quad \varepsilon : \begin{cases} A \to k, \\ n \mapsto 0, \\ s \mapsto 1, \end{cases}$$

endow A with a structure of non commutative and non cocommutative bialgebra.

(3) Prove that the map α defined by

$$\alpha : n \mapsto -n', \ s \mapsto s, \ n' \mapsto n,$$

is an antipode, of order 4.

Exercise 8.4.15 For k a (commutative) field, G a finite group, $s \in G$, S a $kC_G(s)$ module, we denote by $X(s, S)$ the corresponding $D_k G$-module. Prove that $X(s, S)^* \cong X(s^{-1}, S^*)$.

The following exercises follow the lines of [EGNO15, Sect. 8.13]

Exercise 8.4.16 Give a direct and "abstract" proof of Corollary 8.3.20, which uses neither Proposition 8.3.14 nor Proposition 8.3.19.

Hint: Use the equality

$$(c_{Y,X} \otimes \mathrm{Id}_Z)(\mathrm{Id}_X \otimes c_{Z,X}c_{X,Z})(c_{X,Y} \otimes \mathrm{Id}_Z) = c_{Y \otimes Z, X}c_{X, Y \otimes Z}.$$

Exercise 8.4.17 With the notation of Sect. 8.3.2, prove that for all $X, Y \in \mathrm{Irr}(D_K G)$

$$\theta_X \theta_Y s_{X,Y} = \sum_{Z \in \mathrm{Irr}(D_K G)} N_{X,Y}^Z \theta_Z \dim Z.$$

Exercise 8.4.18 We use the notation of Sect. 8.3.2. For $X \in \mathrm{Irr}(D_K G)$, let us denote by H_X the matrix $(N_{X,Y}^Z)_{Y, Z \in \mathrm{Irr}(D_K G)}$.

(1) Prove that for all $W \in \mathrm{Irr}(D_K G)$, $\mathbf{h}_W(X)$ is an eigenvalue of the matrix H_X.

(2) Prove that all the numbers $\dfrac{s_{X,Y}}{\dim X}$ are algebraic integers.

Appendix A
Basics on Finite Groups

A.1 π-Parts of Elements

Let G be a group.

Proposition A.1.1 *Let* $g \in G$ *be an element of order n. Assume* $n = ab$ *where a and b are relatively prime integers.*

(1) *There exist elements* g_a *and* g_b *of G, uniquely defined by the two conditions:*

   (c) $g = g_a g_b = g_b g_a$,
   (o) *the order of* g_a *divides a, and the order of* g_b *divides b.*

(2) *Assume* g_a *and* g_b *as above.*

   (a) *If* $ua + vb = 1$, *we have* $g_a = g^{vb}$ *and* $g_b = g^{ua}$.
   (b) g_a *has order a and* g_b *has order b.*

Proof Assume $ua + vb = 1$.

We have $g = g^{vb}g^{ua} = g^{ua}g^{vb}$; moreover, the order of g^{vb} divides a and the order of g^{ua} divides b, which shows the existence of elements $g_a := g^{vb}$ and $g_b := g^{va}$ satisfying (c) and (o).

Let us now prove the uniqueness. Thus assume that there exist g_a and g_b in G satisfying (c) and (o). We have $g_a = g_a^{ua}g_a^{vb} = g_a^{vb}$, so $g^{vb} = g_a^{vb}g_b^{vb} = g_a$. Similarly, $g^{ua} = g_b$, which establishes the uniqueness.

Finally, the order of g is the lcm of the orders of g_a (a divisor of a) and of g_b (a divisor of b), and since that order is ab it follows that the order of g_a is a and the order of g_b is b. $\square$

Notation A.1.2 Let π be a set of prime numbers. We denote by π' the complement of π, that is, the set of all prime numbers which do not belong to π.

(1) For $n \in \mathbb{N}$, we write $n = n_\pi n_{\pi'}$ where all the prime divisors of n_π belong to π while all the prime divisors of $n_{\pi'}$ belong to π'.

© Springer Nature Singapore Pte Ltd. 2017
M. Broué, *On Characters of Finite Groups*, Mathematical Lectures from Peking University, https://doi.org/10.1007/978-981-10-6878-2

(2) For $g \in G$ of order n, we set

$$g_\pi := g_{n_\pi} \quad \text{and} \quad g_{\pi'} := g_{n_{\pi'}},$$

and we call g_π the π-*component* of g (hence $g_{\pi'}$ is the π'-*component* of g).
(3) If $\pi = \{p\}$ is a singleton, we set $g_p := g_{\{p\}}$ and $g_{p'} := g_{\{p\}'}$. Then g_p and $g_{p'}$ are called respectively the p-component and the p'-component of g.

A.2 Nilpotent Groups

Definition A.2.1 A group G is said to be *nilpotent* if there exists a finite chain of normal subgroups $1 = G_0 \subset G_1 \subset \cdots \subset G_n = G$ such that

$$\text{for } i = 1, \ldots, n, \ G_i/G_{i-1} = Z(G/G_{i-1}).$$

The sequence described in the above definition is called the *ascending central chain*.

Notice that for $i = 1, \ldots, n$, $[G, G_i] \subset G_{i-1}$.

Examples A.2.2

- An abelian group is nilpotent, but the group $\mathfrak{S}_3$ is not nilpotent.
- A direct product of nilpotent groups is nilpotent.
- If p is a prime, any finite p-group is nilpotent.

Indeed, this follows from the following lemma.

Lemma A.2.3 *Let p be a prime, and let G be a finite p-group. Then $Z(G) \neq 1$.*

Proof We let G act on itself by conjugation. The decomposition into the disjoint union of orbits gives

$$G = Z(G) \sqcup \bigsqcup_C C$$

where C runs over the set of non central conjugacy classes. Thus $|G| = |Z(G)| + \sum_C |C|$ and p divides $|Z(G)|$, hence $Z(G) \neq 1$. $\qquad \square$

Let us give a few properties of nilpotent groups.

Proposition A.2.4 *Let G be a nilpotent finite group.*

(1) *If H is a proper subgroup of G, then H is a proper subgroup of its normalizer.*
(2) *If G is nonabelian, there exists an abelian normal subgroup of G not contained in $Z(G)$.*

Proof

(1) Let $1 = G_0 \subset \cdots \subset G_n = G$ be a chain as in Definition A.2.1. Since $H = HG_0 \subset HG_1 \subset \cdots \subset HG_n = G$, there exists $i < n$ such that $G_i \subset H$ and $H \not\subset G_{i+1}$. Then $H \subsetneq HG_{i+1}$, and $H \lhd HG_{i+1}$ since $[H, G_{i+1}] \subset G_i \subset H$.

(2) Since G is nonabelian, G_1 is a proper subgroup of G, hence $G_1 \subsetneq G_2$. Choose $g \in G_2 \setminus G_1$. Then the group $A := G_1 \langle g \rangle$ is abelian and not contained in G_1. It is normal since $[G, A] \subset G_1$. $\qquad\square$

Theorem A.2.5 *Let G be a finite group. The following assertions are equivalent.*

 (i) *G is nilpotent.*
 (ii) *There exists a (finite) family of primes p and of p-subgroups G_p of G such that G is isomorphic to the direct product of the G_p's.*

Proof (ii)$\Rightarrow$(i) is trivial (see Example A.2.2). Let us prove (i)$\Rightarrow$(ii).

1. We first prove that, for every prime p, a Sylow p-subgroup P of G is normal in G.

This follows from Proposition A.2.4, (1) above, and from the following general lemma.

Lemma A.2.6 *Whenever G is a finite group, p a prime number, and P a Sylow p-subgroup of G, then $N_G(N_G(P)) = N_G(P)$.*

Proof of Lemma A.2.6 Indeed, since P is the unique Sylow p-subgroup of $N_G(P)$, P is a characteristic subgroup of $N_G(P)$ (that is, a subgroup which is stable under *any* automorphism of $N_G(P)$). $\qquad\square$

2. Let $p_1, \ldots, p_m$ be the different prime numbers dividing $|G|$ and let $P_1, \ldots, P_m$ be the corresponding Sylow subgroups of G, all normal in G. Thus for all $i \neq j$ we have $[P_i, P_j] \subset P_i \cap P_j$, hence $[P_i, P_j] = 1$. It follows that the map

$$P_1 \times \cdots \times P_m \to P_1 \cdots P_m \subset G , \quad (x_1, \ldots, x_m) \mapsto x_1 \cdots x_m ,$$

is a surjective group morphism. Since the order of its image $P_1 \cdots P_m$ is divisible by the order of each Sylow subgroup of G, we have $P_1 \cdots P_m = G$, and the above map is an isomorphism. $\qquad\square$

A.3   Complements on Sylow Subgroups

The next proposition is known as the *"local characterization"* of *"Sylow subgroups"*.

Proposition A.3.1 *Let P be a p-subgroup of a finite group G. The following assertions are equivalent.*

(i) *P is a Sylow p-subgroup of G.*
(ii) *P is a Sylow p-subgroup of its normalizer $N_G(P)$.*

Proof (i)$\Rightarrow$(ii) is clear. Let us prove (ii)$\Rightarrow$(i). For this purpose, we prove that if there exists a p-subgroup Q of G such that $P \subsetneq Q$, then $P \subsetneq N_Q(P)$. Indeed, since Q is nilpotent, this follows from item (1) of Proposition A.2.4. $\qquad\square$

The next proposition is known as the *"Frattini argument"*.

Proposition A.3.2 (Frattini argument) *Assume that H is a normal subgroup of G. Let p be a prime and let P be a Sylow p-subgroup of H. Then*

$$G = HN_G(P).$$

Proof For $g \in G$, gPg^{-1} is still a Sylow p-subgroup of H. Hence there exists $h \in H$ such that $gPg^{-1} = hPh^{-1}$. Thus $n := h^{-1}g \in N_G(P)$ and $g = hn$. $\qquad\square$

A.4   Schur–Zassenhaus Theorem

A proof of the following statement may be found in [Gor80], Chap. 6, Theorem 2.1.

Theorem A.4.1 (Schur–Zassenhaus) *Assume that N is a normal subgroup of a finite group G such that $|N|$ and $|G/N|$ are relatively prime.*

(1) *There exists a subgroup H of G such that $G = NH$ and $N \cap H = 1$, i.e., $G \simeq N \rtimes H$.*
(2) *If either N or G/N is solvable, then the subgroups H of G such that $G = NH$ and $N \cap H = 1$ form a single conjugacy class of subgroups of G.*

Remark A.4.2 A celebrated (very deep and difficult) theorem of Feit and Thompson (W. Feit and J.G. Thompson, *Solvability of groups of odd order*, Pacific J. Math. 13 (1963), 775–1029) proves that any group of odd order is solvable.

Hence, by this theorem, we see that the assumption of solvability can be dropped in the second assertion of the Schur–Zassenhaus Theorem.

Appendix B
Assumed Results on Galois Theory

For the contents of this appendix, see for example [Ste03].

All fields considered here are commutative and have characteristic zero.

B.1 Galois Extensions

For the following fundamental property, the reader may also look at [Bro13], Proposition 1.78.

Proposition B.1.1 *Let K' be a finite extension of a field K. The following assertions are equivalent.*

(i) *There exists $P(X) \in K[X]$ such that K' is the splitting field of $P(X)$ over K.*
(ii) *Whenever $Q(X)$ is an irreducible element of $K[X]$ which has at least one root in K', then $Q(X)$ splits into a product of degree one factors over K'.*

Definition B.1.2

(1) A finite field extension K'/K is called a *Galois extension* if it satisfies properties (i) and (ii) as above.
(2) The Galois group $\mathrm{Gal}(K'/K)$ of such an extension is the set of automorphisms of K' which induce the identity on K.

Examples B.1.3

(1) Let $\alpha := \sqrt[3]{2} \in \mathbb{R}$. Then the extension $\mathbb{Q}(\alpha)/\mathbb{Q}$ is not Galois.
   Indeed, two of the roots of $X^3 - 2$ (which is irreducible in $\mathbb{Q}[X]$) do not belong to $\mathbb{Q}(\alpha)$.
(2) Let $n \in \mathbb{N}$, $n \geq 1$, and let ζ be a root of unity of order n in an extension of K. Then the extension $K(\zeta)/K$ is Galois.
   Indeed, since any n-th root of 1 is a power of ζ, the field $K(\zeta)$ is the splitting field of $X^n - 1$ over K.

© Springer Nature Singapore Pte Ltd. 2017
M. Broué, *On Characters of Finite Groups*, Mathematical Lectures from Peking
University, https://doi.org/10.1007/978-981-10-6878-2

(3) The Galois group $\text{Gal}(K(\zeta)/K)$ is isomorphic to a subgroup of the multiplicative group $(\mathbb{Z}/n\mathbb{Z})^{\times}$.

Indeed, if $\sigma \in \text{Gal}(K(\zeta)/K)$, then $\sigma(\zeta)$ is again a root of unity of order n, hence there exists $a(\sigma) \in (\mathbb{Z}/n\mathbb{Z})^{\times}$ such that $g(\zeta) = \zeta^{a(\sigma)}$, and the map

$$\text{Gal}(K(\zeta)/K) \to (\mathbb{Z}/n\mathbb{Z})^{\times} \;, \quad \sigma \mapsto a(\sigma),$$

is clearly an injective group morphism.

B.2 Main Theorem

Theorem B.2.1 *Let $\phi : K_1 \xrightarrow{\sim} K$ be a field isomorphism.*

Let L/K_1 be a finite extension, and let K'/K be a Galois extension such that K' contains the roots of $\phi(P(X))$ where $P(X)$ runs over the set of minimal polynomials over K_1 of elements of L.

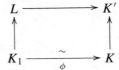

Then there exist exactly $[L : K_1]$ morphisms $L \to K'$ extending ϕ.

Sketch of proof This is essentially, for example, the proof of Proposition 2.246 in [Bro13]. □

Corollary B.2.2 *Let L/K be a finite extension and let K' be a Galois extension of K which contains L.*

Let $\text{Mor}_K(L, K')$ denote the set of morphisms $L \to K'$ inducing the identity on K.

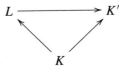

(1) $|\text{Mor}_K(L, K')| = [L : K]$.

(2) *Let $x \in L$. Then*

$$x \in K \iff \forall \sigma \in \text{Mor}_K(L, K'), \; \sigma(x) = x.$$

Proof

(1) is a particular case of the preceding theorem.

In order to prove (2), it is enough to prove that if $x \in L$, $x \notin K$, there exists $\sigma \in \text{Mor}_K(L, K')$ such that $\sigma(x) \neq x$.

So assume $x \in L \setminus K$. The irreducible polynomial of x over L has degree at least 2, and splits into degree 1 polynomials over K', hence has a root $y \neq x$ in K' (since we are in characteristic zero). The map $x \mapsto y$ induces an isomorphism $K(x) \xrightarrow{\sim} K(y)$ by Theorem B.2.1. That isomorphism extends to $[L : K(x)]$ morphisms $L \to K'$, and, for such morphisms σ, we do have $\sigma(x) \neq x$.   □

The following corollary is an immediate consequence of the preceding one.

Corollary B.2.3 *Let K'/K be a Galois extension.*

(1) $|\mathrm{Gal}(K'/K)| = [K' : K]$.
(2) $\mathrm{Fix}^{\mathrm{Gal}(K'/K)}(K') = K$.

Example B.2.4 Let $\zeta_n := e^{2\pi i/n}$.
  Since the cyclotomic polynomial $\Phi_n(X)$ (see e.g. [Bro13, 1.1.4.3]) has degree $\varphi(n) = |(\mathbb{Z}/n\mathbb{Z})^\times|$, and since it is irreducible over $\mathbb{Q}$ ([Bro13, Theorem 1.171]), we see that

$$[\mathbb{Q}(\zeta_n)/\mathbb{Q}] = |(\mathbb{Z}/n\mathbb{Z})^\times|,$$

and so the morphism $\mathrm{Gal}(\mathbb{Q}(\zeta_n)/\mathbb{Q}) \to (\mathbb{Z}/n\mathbb{Z})^\times$ described in Example B.1.3, (2), is an isomorphism.

Exercise B.2.5 Let $\zeta := \exp(\frac{2i\pi}{5})$, and let $P(X) := X^4 + X^3 + X^2 + X + 1$ (see Exercise 3.7.21).
  Let ϕ be the golden ratio, characterized by the conditions

$$\phi^2 = \phi + 1 \quad \text{and} \quad \phi > 0.$$

(1) Prove that
$$\zeta^2 + \zeta^3 = -\phi \quad \text{and} \quad \zeta + \zeta^4 = \phi - 1.$$

(2) Prove that $P(X)$ has the following decomposition over the field $\mathbb{Q}(\phi)[X]$:

$$P(X) = (X^2 + \phi X + 1)(X^2 + (1 - \phi)X + 1).$$

(3) Prove that $P(X)$ is irreducible over $\mathbb{Q}(2^{1/5})$ and deduce that the field $\mathbb{Q}(2^{1/5}, \zeta)$ has degree 20 over $\mathbb{Q}$.
(4) Prove that $\mathbb{Q}(2^{1/5}, \zeta)/\mathbb{Q}$ is a Galois extension and describe its Galois group $\mathrm{Gal}(\mathbb{Q}(2^{1/5}, \zeta)/\mathbb{Q})$.

B.3   Minimal and Characteristic Polynomials, Morphisms

Let L/K be a finite extension and let $x \in L$.
  The $[K(x) : K]$ elements $\sigma(x)_{\sigma \in \mathrm{Mor}_K(K(x), K')}$ (see Theorem B.2.1) are called the *conjugates* of x over K.

We denote by

- $M_x(X)$ the minimal polynomial of x over L, so that $K(x) \simeq K[X]/(M_x(X))$,
- $m_{L/K,x}$ the K-linear endomorphism of L defined by the multiplication by x, and

  - $\Gamma_{L/K,x}(X)$ the characteristic polynomial of $m_{L/K,x}$,
  - $\mathrm{Tr}_{L/K}(x)$ the trace of $m_{L/K,x}$,
  - $N_{L/K}(x)$ the determinant of $m_{L/K,x}$, called the *norm* of x relative to the extension L/K.

The following proposition is essentially Proposition 2.247 in [Bro13].

Proposition B.3.1 *Let L/K be a finite extension and let K' be a Galois extension of K which contains L.*

Let $x \in K$.

We denote by $\mathrm{Gal}(K'/K)_x$ the fixator of x in $\mathrm{Gal}(K'/K)$. We call conjugates *of x the images of x under* $\mathrm{Gal}(K'/K)$.

(1) $M_x(X) = \prod_{\sigma \in \mathrm{Mor}_K(K(x),K')}(X - \sigma(x))$,

(2) $M_x(X) = \prod_{\sigma \in \mathrm{Gal}(K'/K)/\mathrm{Gal}(K'/K)_x}(X - \sigma(x))$, *that is, the roots of $M_x(X)$ are the distinct conjugates of x.*

(3) $\Gamma_{L/K,x}(X) = M_x(X)^{[L:K(x)]} = \prod_{\sigma \in \mathrm{Mor}_K(L,K')}(X - \sigma(x))$, *and in particular*

  (a) $\mathrm{Tr}_{L/K}(x) = \sum_{\sigma \in \mathrm{Mor}_K(L,K')} \sigma(x)$,

  (b) $N_{L/K}(x) = \prod_{\sigma \in \mathrm{Mor}_K(L,K')} \sigma(x)$.

Appendix C
Integral Elements

All throughout this section, A is assumed to be a subring of a commutative ring B.

C.1 Definition, Integral Closure

Definition C.1.1 Let $x \in B$. We say that x is *integral over A* if there exists a *monic* polynomial $P(X) \in A[X]$ such that $P(x) = 0$.

Examples C.1.2 Assume $A = \mathbb{Z}$ and $B = \mathbb{C}$. In that case, an element which is integral over $\mathbb{Z}$ is called an *algebraic integer*.

(1) The elements of $\mathbb{Z}$ are integral over $\mathbb{Z}$.
(2) All roots of unity of $\mathbb{C}$, all elements $\sqrt[m]{n}$ for $m, n \in \mathbb{N}$, $m, n \geq 1$, all elements $(1 + \sqrt{d})/2$ for $d \in \mathbb{Z}$ and $d \equiv 1 \mod 4$ (thus in particular the Golden Ratio $(1 + \sqrt{5})/2$) are algebraic integers.

Exercise C.1.3 Prove that $x \in \mathbb{Q}$ is integral over $\mathbb{Z}$ if and only if $x \in \mathbb{Z}$.

Proposition C.1.4 *Let $x \in B$. The following assertions are equivalent:*

(i) *the element x is integral over A,*
(ii) *the subring $A[x]$ of B generated by A and x is a finitely generated A–module,*
(iii) *there exists a subring A' of B, containing x, which is a finitely generated A–module.*

Proof
(i)$\Rightarrow$(ii): Assume that $x^r - \lambda_{r-1}x^{r-1} - \cdots - \lambda_1 x - \lambda_0 = 0$. We shall prove that $A[x]$ is generated by $\{1, x, \ldots, x^{r-1}\}$.

Since $A[x]$ is generated by $\{x^n\}_{n\in\mathbb{N}}$, it suffices to prove that, for all $n \geq 0$, x^{r+n} is a linear combination of $\{1, x, \ldots, x^{r-1}\}$. It is clear for $n = 0$. The proof by induction on n is easy.

(ii)$\Rightarrow$(iii): trivial.

© Springer Nature Singapore Pte Ltd. 2017
M. Broué, *On Characters of Finite Groups*, Mathematical Lectures from Peking University, https://doi.org/10.1007/978-981-10-6878-2

(iii)$\Rightarrow$(i): Assume that $\{x_1, \ldots, x_m\}$ is a generating system for A'. Since $x \in A'$, there are $\lambda_{i,j} \in A$ ($1 \leq i, j \leq m$) such that for all i, $x x_i = \sum_{j=1}^{m} \lambda_{i,j} x_j$. Let Λ be the $m \times m$ square matrix with entries $\lambda_{i,j} \in A$ ($1 \leq i, j \leq m$). The above equations give

$$(x 1_m - \Lambda) \begin{pmatrix} x_1 \\ \vdots \\ x_m \end{pmatrix} = \begin{pmatrix} 0 \\ \vdots \\ 0 \end{pmatrix}.$$

Multiplying to the left by ${}^t\mathrm{Com}(x 1_m - \Lambda)$ gives then

$$\det(x 1_m - \Lambda) \begin{pmatrix} x_1 \\ \vdots \\ x_m \end{pmatrix} = \begin{pmatrix} 0 \\ \vdots \\ 0 \end{pmatrix},$$

so

$$\det(x 1_m - \Lambda) A' = 0 \quad \text{hence} \quad \det(x 1_m - \Lambda) = 0.$$

Now $\det(X 1_m - \Lambda)$ is a monic element of $A[X]$, which proves (i). $\square$

Corollary C.1.5 *The set of elements of B which are integral over A is a subring of B which contains A.*

Proof We must prove that whenever x and y are two elements of B which are integral over A, then $x + y$ and xy are integral over A. This follows from the fact that all the elements of the ring $A[x, y]$ are integral over A, which we prove now.

By Proposition C.1.4, it suffices to prove that $A[x, y]$ is a finitely generated A-module. This follows from the following lemma (which generalizes part of a known result about fields extensions).

Lemma C.1.6 *Assume that $A = A_0 \subset A_1 \subset \cdots \subset A_m$ is a tower of commutative rings, where (for all $i = 0, \ldots, m - 1$) A_i is a subring of A_{i+1}. If, for all $i = 0, \ldots, m - 1$, A_{i+1} is a finitely generated A_i-module, then A_m is a finitely generated A-module.*

Proof of Lemma C.1.6 We sketch a proof in the case where $m = 2$ (which is sufficient). Let us set $B := A_1$ and $C := A_2$. Let $\{s_1, \ldots, s_m\}$ be a generating system of B as an A-module, and let $\{t_1, \ldots, t_n\}$ be a generating system of C as a B-module. Then it can be checked (do it!) that $\{s_i t_j\}_{1 \leq i \leq m, 1 \leq j \leq n}$ is a generating system of C as an A-module. $\square$

Definition C.1.7

(1) The ring of elements of B which are integral over A is called *the integral closure* of A in B.
(2) If $A = \mathbb{Z}$ and $B = \mathbb{C}$, the integral closure of $\mathbb{Z}$ in $\mathbb{C}$ (the ring of all algebraic integers) is denoted by $\bar{\mathbb{Z}}$.

C.2   A Few Properties

C.2.1   *Minimal Polynomial of an Integral Element*

The following property shows that if an element is integral over $\mathbb{Z}$, it is the root of a monic element of $\mathbb{Z}[X]$ which is *irreducible* over $\mathbb{Q}$.

Proposition C.2.1 *Let $x \in \mathbb{C}$ be an algebraic integer, and let $M_x(X)$ denote its monic minimal polynomial over $\mathbb{Q}$. Then*

$$M_x(X) \in \mathbb{Z}[X].$$

Sketch of proof Let $P(X) \in \mathbb{Z}[X]$ be a monic polynomial such that $P(x) = 0$. Whenever $\sigma : \mathbb{Q}(x) \to K'$ is a field morphism into a Galois extension K' of $\mathbb{Q}$ (note that σ induces the identity on $\mathbb{Q}$), then $\sigma(x)$ is also a root of $P(X)$, hence is an algebraic integer. By Proposition B.3.1, we know that $M_x(X) = \prod_{\sigma \in \mathrm{Mor}_{\mathbb{Q}}(\mathbb{Q}(x), K')}(X - \sigma(x))$, from which it follows that the coefficients of $M_x(X)$, which are symmetric polynomials evaluated at the family $(\sigma(x))_{\sigma \in \mathrm{Mor}_{\mathbb{Q}}(\mathbb{Q}(x), K')}$,

- are fixed by all $\sigma \in \mathrm{Mor}_{\mathbb{Q}}(\mathbb{Q}(x), K')$, hence (see Corollary B.2.2, (2)) belong to $\mathbb{Q}$,
- are algebraic integers (by Corollary C.1.5),

hence belong to $\mathbb{Z}$ (by Exercise C.1.3).   □

Remark C.2.2 The reader may state various generalizations of the previous proposition (for example, replacing $\mathbb{Z}$ by any factorial domain).

C.2.2   *On Fields of Fractions and Integrality*

We conclude by a lemma which is related to Proposition 7.2.6.

Lemma C.2.3 *Let A be a subring of an integral domain B. Let K (resp. L) be the field of fractions of A (resp. B).*
   Assume that B is integral over A.

(1) *L is algebraic over K.*
(2) *Each element of L may be written b/a where $b \in B$ and $a \in A$.*
(3) *Every system of generators of B as an A-module is also a system of generators of L as a K-vector space.*

Proof of Lemma C.2.3
(1) Since L is generated over K by the elements of B which are all algebraic over K, L is algebraic over K.

(2) Let $b \in B$, $b \notin A$. There exist $d \geq 2$ and $a_0, \ldots, a_{d-1} \in A$ such that $b^d + a_{d-1}b^{d-1} + \cdots + a_0 = 0$. Let e be the smallest integer such that $a_e \neq 0$. Since $b \notin A$, $e \leq d - 1$, and $b^{-1} = -a_e^{-1}\left(b^{d-e-1} + a_{d-1}b^{d-e-2} + \cdots + a_{e+1}\right)$, showing that $b^{-1} = b'/a$ for some $b' \in B$ and $a \in A$.

(3) Assume that $B = Ab_1 + \cdots + Ab_s$. Then every element b/a of L may be written $b/a = (a_1/a)b_1 + \cdots + (a_s/a)b_s$. $\qquad\square$

Appendix D
Noetherian Rings and Modules

D.1 Noetherian Modules

Theorem D.1.1 *Let R be a commutative ring and let M be an R-module. The following conditions are equivalent.*

  (i) *Every nonempty family of submodules of M has a maximal element.*
 (ii) *Every increasing (ascending) sequence of submodules $M_0 \subset M_1 \subset \cdots \subset M_n \subset \cdots$ of M is stationary.*
(iii) *Every submodule of M is finitely generated.*

The following definition is in honor of Emmy Noether.

Definition D.1.2 A module satisfying the above equivalent conditions is said to be *Noetherian.*

Proof of Theorem D.1.1
(i)$\Rightarrow$(iii) Let N be a submodule of M, and let $\mathcal{F}(N)$ be the family of all finitely generated submodules of N. Since $\mathcal{F}(N)$ contains the trivial module 0, it is nonempty hence it has a maximal element, say N'. Let us show that $N' = N$. Let $x \in N$. The module $N' + Rx$ is finitely generated, hence $N' + Rx = N'$, showing that $x \in N'$.
(iii)$\Rightarrow$(ii) Let $(M_n)_{n \geq 0}$ be an increasing sequence of submodules of M. It is easy to check that $M_\infty := \bigcup_{n \geq 0} M_n$ is a submodule of M. Since it is finitely generated, there exists $m \in \mathbb{N}$ such that M_m contains a set of generators, which implies $M_\infty = M_m$, hence that the sequence $(M_n)_{n \geq 0}$ is stationary.
(ii)$\Rightarrow$(i) The proof relies on the following lemma. Note that this lemma uses the axiom of choice (where?).

Lemma D.1.3 *Let Ω be a (partially) ordered set. The following assertions are equivalent.*

  (i) *Every nonempty subset of Ω has a maximal element.*
 (ii) *Every increasing sequence $(\omega_n)_{n \geq 0}$ of elements of Ω is stationary.*

© Springer Nature Singapore Pte Ltd. 2017
M. Broué, *On Characters of Finite Groups*, Mathematical Lectures from Peking
University, https://doi.org/10.1007/978-981-10-6878-2

Proof of Lemma D.1.3

(i)$\Rightarrow$(ii): Let $(\omega_n)_{n\geq 0}$ be an increasing sequence. If ω_m is a maximal element, we have $\omega_n = \omega_m$ for all $n \geq m$, which shows that the sequence is stationary.

(ii)$\Rightarrow$(i): Assume (i) is false, and let Ω' be a nonempty subset of Ω which has no maximal element. Let $\omega_0 \in \Omega'$. We can then build by induction a strictly increasing sequence $(\omega_n)_{n\geq 0}$ in Ω' as follows: Assume ω_n known. Since ω_n is not maximal in Ω', we may pick $\omega_{n+1} \in \Omega'$ such that $\omega_{n+1} > \omega_n$.

Proposition D.1.4 *Let* $0 \to M' \xrightarrow{\iota} M \xrightarrow{\pi} M'' \to 0$ *be a short exact sequence of R-modules. Then the following assertions are equivalent:*

(i) *M is Noetherian,*

(ii) *M' and M'' are Noetherian.*

Proof (i)$\Rightarrow$(ii): Any increasing sequence of submodules of M' (or M'') gives rise to an increasing sequence of submodules of M, hence is stationary.

(ii)$\Rightarrow$(i): Let us prove that any submodule N of M is finitely generated. By assumption, the submodule $\iota^{-1}(N)$ of M' and the submodule $\pi(N)$ of M'' are finitely generated. Let $(x_1, \ldots, x_m)$ be the image under ι of a set of generators of $\iota^{-1}(N)$, and let $(y_1, \ldots, y_n)$ be a set of preimages under π of generators of $\pi(N)$. The reader will check as an exercise that $(x_1, \ldots, x_m, y_1, \ldots, y_n)$ is a set of generators of N. $\square$

Corollary D.1.5 *If* $M_1, M_2, \ldots, M_r$ *are Noetherian R-modules, so is* $\bigoplus_{n=1}^{r} M_n$.

Proof Applying Proposition D.1.4 to the short exact sequence $0 \to M_1 \to M_1 \oplus M_2 \to M_2 \to 0$ gives that $M_1 \oplus M_2$ is Noetherian, and now an easy induction on r proves the claim. $\square$

D.2 Noetherian Rings

Definition D.2.1 A ring is a *Noetherian ring* if it is Noetherian as a module over itself.

⚠ Let k be a field. Then the polynomial ring $k[(X_n)_{n\geq 0}]$ in a countable set of indeterminates is *not* Noetherian. That ring is a subring of its field of fractions $k((X_n)_{n\geq 0})$, hence a subring of a Noetherian ring needs not be Noetherian.

Proposition D.2.2 *Let R be a commutative Noetherian ring.*

(1) *If* $\mathfrak{a}$ *is an ideal of R, the ring* $R/\mathfrak{a}$ *is Noetherian.*

(2) *Any finitely generated R-module is Noetherian.*

(3) *If R is a subring of a ring T which is a finitely generated R-module, then T is a Noetherian ring.*

Proof
(1) $R/\mathfrak{a}$ is Noetherian as an R-module by Proposition D.1.4, hence is Noetherian as an $R/\mathfrak{a}$-module, that is, is a Noetherian ring.
(2) Assume M is an image of R^m for some integer m. Since R^m is a Noetherian R-module by Corollary D.1.5, M is Noetherian by Proposition D.1.4.
(3) T is a Noetherian R-module by (2) above, hence *a fortiori* a Noetherian T-module, hence a Noetherian ring.   □

D.3   Hilbert's Basis Theorem

Theorem D.3.1 (Hilbert's Basis Theorem) *Let R be a Noetherian ring. Then the polynomial ring $R[X_1, \ldots, X_r]$ in a finite number of indeterminates is Noetherian.*

Proof It suffices to prove that $R[X]$ is Noetherian.

Let $\mathfrak{A}$ be an ideal of $R[X]$. We shall prove that $\mathfrak{A}$ is finitely generated.

The leading coefficients of the elements of $\mathfrak{A}$ form an ideal $\mathfrak{a}$ in R. That ideal is finitely generated since R is Noetherian. Let $a_1, \ldots, a_m$ be a set of generators of $\mathfrak{a}$. Let $P_1(X), \ldots, P_m(X) \in \mathfrak{A}$ whose leading coefficients are respectively $a_1, \ldots, a_m$ and let us denote by $\mathfrak{B}$ the ideal of $R[X]$ generated by $P_1(X), \ldots, P_m(X)$.

For $1 \leq i \leq m$, we set $d_i := \deg P_i(X)$, and $r := \max\{d_1, \ldots, d_m\}$. Let $R[X]_r$ be the R-submodule of $R[X]$ generated by $\{1, X, \ldots, X^{r-1}\}$. In order to prove that $\mathfrak{A}$ is finitely generated, we shall prove that

$$\mathfrak{A} = (\mathfrak{A} \cap R[X]_r) + \mathfrak{B}. \tag{D.3.2}$$

Since $\mathfrak{B}$ is finitely generated by definition, and $\mathfrak{A} \cap R[X]_r$ is finitely generated as an R-module (since it is an R-submodule of the finitely generated R-module $R[X]_r$ and R is Noetherian), that will prove the result.

Let $P(X) \in \mathfrak{A}$. Let $d := \deg P(X)$. We may assume that $d > r$. Let a be its leading coefficient. We have $a = \lambda_1 a_1 + \cdots + \lambda_m a_m$ for some $\lambda_i \in R$. Then the polynomial

$$P(X) - \sum_{i=1}^{m} \lambda_i X^{d-d_i} P_i(X)$$

has degree strictly smaller than d. Repeating that operation, we get $P(X)$ as the sum of an element of $R[X]_r$ and an element of $\mathfrak{B}$, thus proving the announced equality (D.3.2).   □

Corollary D.3.3 *If R is a Noetherian ring, any finitely generated R-algebra is a Noetherian ring.*

Proof Indeed, it is an immediate consequence of Theorem D.3.1 and of Proposition D.2.2, (1).   □

Appendix E
The Language of Categories and Functors

E.1 General Definitions

We briefly introduce (or recall) some basic notation and definitions about categories. For more details we refer the reader to [Mac71].

E.1.1 Categories and Functors

Categories.

Definition E.1.1 A *category* consists of the following three mathematical entities:

- A class $\mathrm{Ob}(\mathfrak{C})$, whose elements are called *objects* (for X an object, we write $X \in \mathrm{Ob}(\mathfrak{C})$, or even $X \in \mathfrak{C}$),
- for each pair of objects X and X', a set $\mathrm{Mor}_{\mathfrak{C}}(X, X')$ (an element $f \in \mathrm{Mor}_{\mathfrak{C}}(X, X')$ is then called a *morphism* with source X and target X' and denoted $f : X \to X'$),
- for each triple of objects X, X', X'', a map called the composition:

$$\begin{cases} \mathrm{Mor}_{\mathfrak{C}}(X', X'') \times \mathrm{Mor}_{\mathfrak{C}}(X, X') \longrightarrow \mathrm{Mor}_{\mathfrak{C}}(X, X'') \\ (g, f) \mapsto g \cdot f\,, \end{cases}$$

such that

(1) $(h \cdot g) \cdot f = h \cdot (g \cdot f)$,
(2) whenever $X \in \mathfrak{C}$, there is an element $\mathrm{Id}_X \in \mathrm{Mor}_{\mathfrak{C}}(X, X)$ such that, for every morphism $f : X \to X'$, we have $f \cdot \mathrm{Id}_X = \mathrm{Id}_{X'} \cdot f$.

Let us give some definitions related to properties of morphisms.
A morphism $f : X \to X'$ is

© Springer Nature Singapore Pte Ltd. 2017
M. Broué, *On Characters of Finite Groups*, Mathematical Lectures from Peking
University, https://doi.org/10.1007/978-981-10-6878-2

- a *monomorphism* (or monic) if $f \cdot g_1 = f \cdot g_2$ implies $g_1 = g_2$ for all morphisms $g_1, g_2 : X'' \to X$,
- an *epimorphism* (or epic) if $g_1 \cdot f = g_2 \cdot f$ implies $g_1 = g_2$ for all morphisms $g_1, g_2 : X' \to X''$,
- an *isomorphism* if there exists a morphism $g : X' \to X$ with $f \cdot g = \mathrm{Id}_{X'}$ and $g \cdot f = \mathrm{Id}_X$,
- an *endomorphism* if $X' = X$ ($\mathrm{End}_{\mathfrak{C}}(X)$ denotes the set of endomorphisms of X,
- an *automorphism* is both an endomorphism and an isomorphism. ($\mathrm{Aut}(X)$ denotes the group of automorphisms of X).

A *full subcategory* $\mathfrak{C}'$ of a category $\mathfrak{C}$ is a category where

- the objects of $\mathfrak{C}'$ are some objects of $\mathfrak{C}$,
- for X and X' objects of $\mathfrak{C}'$, we have $\mathrm{Mor}_{\mathfrak{C}}(X, X') = \mathrm{Mor}_{\mathfrak{C}'}(X, X')$.

Example E.1.2 Let k be a field and let A be a finite dimensional k-algebra. We denote by $_A\mathbf{mod}$ the category whose objects are the finitely generated A–modules and where

$$\mathrm{Hom}_{_A\mathbf{mod}}(X, X') := \mathrm{Hom}_A(X, X').$$

Exercise E.1.3 Prove that the monomorphisms (resp. epimorphisms) of $_A\mathbf{mod}$ are the injective (resp. surjective) homomorphisms.

The *opposite category* $\mathfrak{C}^{\mathrm{op}}$ of a category $\mathfrak{C}$ is the category where

- $\mathrm{Ob}(\mathfrak{C}^{\mathrm{op}}) := \mathrm{Ob}(\mathfrak{C})$,
- $\mathrm{Mor}_{\mathfrak{C}^{\mathrm{op}}}(X, X') := \mathrm{Mor}_{\mathfrak{C}}(X', X)$,
- for $f \in \mathrm{Mor}_{\mathfrak{C}^{\mathrm{op}}}(X, X')$ and $g \in \mathrm{Mor}_{\mathfrak{C}^{\mathrm{op}}}(X', X'')$, $(g.f)_{\mathfrak{C}^{\mathrm{op}}} := (f.g)_{\mathfrak{C}}$.

Functors.

Definition E.1.4 Let $\mathfrak{C}$ and $\mathfrak{D}$ be two categories. A *(covariant) functor* $F : \mathfrak{C} \to \mathfrak{D}$

- associates to each $X \in \mathrm{Ob}(\mathfrak{C})$ an object $F(X) \in \mathrm{Ob}(\mathfrak{D})$,
- for each pair (X, X') of objects of $\mathfrak{C}$ it defines a map

$$F : \mathrm{Mor}_{\mathfrak{C}}(X, X') \to \mathrm{Mor}_{\mathfrak{D}}(F(X), F(X'))$$

such that

(1) whenever $X \in \mathfrak{C}$, $F(\mathrm{Id}_X) = \mathrm{Id}_{F(X)}$,
(2) whenever $f : X \to X'$ and $g : X' \to X''$, then $F(g.f) = F(g).F(f)$.

A *contravariant functor* $F : \mathfrak{C} \to \mathfrak{D}$ is a (covariant) functor from $\mathfrak{C}^{\mathrm{op}}$ to $\mathfrak{D}$.

The essential image of F is the full subcategory of $\mathfrak{D}$ whose objects are the objects of $\mathfrak{D}$ isomorphic to objects of the image of F.

We say that the functor F is

- *faithful* if $\text{Mor}_{\mathfrak{C}}(X, X') \to \text{Mor}_{\mathfrak{D}}(F(X), F(X'))$ is injective for all $X, X' \in \mathfrak{C}$,
- *full* if $\text{Mor}_{\mathfrak{C}}(X, X') \to \text{Mor}_{\mathfrak{D}}(F(X), F(X'))$ is surjective for all $X, X' \in \mathfrak{C}$,
- *fully faithful* if it is full and faithful,
- *essentially surjective* if the essential image of F is $\mathfrak{D}$.

Definition E.1.5 Let $F, F' : \mathfrak{C} \to \mathfrak{D}$ be two functors. A morphism $\varepsilon : F \to F'$

- associates to each object X of $\mathfrak{C}$ a morphism

$$\varepsilon_X : F(X) \to F'(X),$$

- such that, whenever $f : X \to X'$ is a morphism in $\mathfrak{C}$, the following diagram is commutative

$$
\begin{array}{ccc}
F(X) & \xrightarrow{\varepsilon_X} & F'(X) \\
{\scriptstyle F(f)}\downarrow & & \downarrow{\scriptstyle F'(f)} \\
F(X') & \xrightarrow{\varepsilon_{X'}} & F'(X')
\end{array}
$$

We say that a morphism $\varepsilon : F \to F'$ is an *isomorphism* if, for all object X of $\mathfrak{C}$, $\varepsilon_X : F(X) \to F'(X)$ is an isomorphism. The functors F and F' are then said to be isomorphic and we write $F \simeq F'$.

Example E.1.6 Let A be a k-algebra. Whenever X is an A–module, the k–module $\text{Hom}_A(A, X)$ is endowed with a natural structure of A–module defined by

$$(a\varphi)(b) := \varphi(ba) \quad \text{for } \varphi \in \text{Hom}_A(A, X), \, a, b \in A.$$

It defines a functor $\text{Hom}_A(A, \cdot)$, which is isomorphic to the functor identity.

Definition E.1.7 We say that a functor $F : \mathfrak{C} \to \mathfrak{D}$ is an *equivalence of categories* if there exists a functor $F' : \mathfrak{D} \to \mathfrak{C}$ such that $F.F' \simeq \text{Id}_{\mathfrak{D}}$ and $F'.F \simeq \text{Id}_{\mathfrak{C}}$.

The proofs of the following two propositions are left to the reader.

Proposition E.1.8 *A functor $F : \mathfrak{C} \to \mathfrak{D}$ is an equivalence of categories if and only if it is fully faithful and essentially surjective.*

Proposition E.1.9 *Assume that the functor $F : \mathfrak{C} \to \mathfrak{D}$ is an equivalence of categories. Then:*

(1) *Whenever $X, X' \in \mathfrak{C}$, then F induces a bijection*

$$\text{Hom}_{\mathfrak{C}}(X, X') \xrightarrow{\sim} \text{Hom}_{\mathfrak{D}}(F(X), F(X')).$$

(2) *The image under F of a monomorphism (resp. an epimorphism) is a monomorphism (resp. an epimorphism).*

E.2   *k*–Linear and Abelian Categories

E.2.1   *k*–Linear Categories

Definition E.2.1 Let k be a field. A category $\mathfrak{A}$ is said to be *k-linear* if the following conditions are satisfied.

(1) • for each pair of objects X and X', $\mathrm{Mor}_{\mathfrak{A}}(X, X')$ is a k-vector space,
   • for each triple of objects X, X', X'', the composition is k-bilinear,

(2) there is a zero object 0, i.e., an object such that for all object X, both $\mathrm{Mor}_{\mathfrak{A}}(0, X)$ and $\mathrm{Mor}_{\mathfrak{A}}(X, 0)$ have a single element ("the 0 morphism"),

(3) every pair of objects X, $X' \in \mathfrak{A}$ admits

   (a) a product, i.e., an object $X \Pi X'$ endowed with morphisms

$$\mathrm{pr}_X : X\Pi X' \to X \quad \text{and} \quad \mathrm{pr}_{X'} : X\Pi X' \to X'$$

   such that the map

$$\mathrm{Mor}_{\mathfrak{A}}(Y, X\Pi X') \longrightarrow \mathrm{Mor}_{\mathfrak{A}}(Y, X) \times \mathrm{Mor}_{\mathfrak{A}}(Y, X')$$
$$\varphi \longmapsto (\mathrm{pr}_X.\varphi, \ \mathrm{pr}_{X'}.\varphi)$$

   is a bijection:

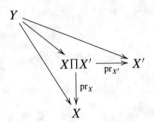

   (b) a coproduct, i.e., an object $X \amalg X'$ endowed with morphisms

$$i_X : X \to X\amalg X' \quad \text{and} \quad i_{X'} : X' \to X\amalg X'$$

   such that the map

$$\mathrm{Mor}_{\mathfrak{A}}(X, Y) \times \mathrm{Mor}_{\mathfrak{A}}(X', Y) \longrightarrow \mathrm{Mor}_{\mathfrak{A}}(X\amalg X', Y)$$
$$\varphi \longmapsto (\varphi.i_X, \ \varphi.i_{X'})$$

is a bijection:

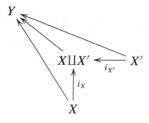

Remark E.2.2 A k–linear category $\mathfrak{A}$ with a single object X_0 is defined by the k–algebra $A := \mathrm{End}_{\mathfrak{A}}(X_0)$.

Exercise E.2.3 Given products $(X \sqcap X', \mathrm{pr}_X, \mathrm{pr}_{X'})$ (resp. coproducts $(X \sqcup X', i_X, i_{X'})$),

(1) prove that there exist coproducts $(X \sqcup X', i_X, i_{X'})$ (resp. products $(X \sqcap X', \mathrm{pr}_X, \mathrm{pr}_{X'})$), and that
$$\begin{cases} i_X.\mathrm{pr}_X + i_{X'}.\mathrm{pr}_{X'} = \mathrm{Id}_Y\,, \\ \mathrm{pr}_X.i_X = \mathrm{Id}_X\,,\ \mathrm{pr}_{X'}.i_{X'} = \mathrm{Id}_{X'}\,, \\ \mathrm{pr}_X.i_{X'} = 0\,,\ \mathrm{pr}_{X'}.i_X = 0\,, \end{cases}$$

(2) there are natural isomorphisms

$$X \sqcap X' \xrightarrow{\sim} X \sqcup X'\,.$$

Definition E.2.4 For X and X' two objects of a k–linear category $\mathfrak{A}$, we say that X' is a summand of X and we write $X' \mid X$ if there exists an object X'' and an isomorphism $X \xrightarrow{\sim} X' \sqcup X''$.

E.2.2   Abelian k-Linear Categories

Throughout this section, objects and morphisms are those of a k–linear category $\mathfrak{A}$.

Kernel, cokernel, image, coimage

- A *kernel* of a morphism $f : X \to Y$, denoted $\ker(f)$, is a pair (X', ι) where $X' \in \mathfrak{A}$ and $\iota : X' \to X$, such that $f\iota = 0$, and given another pair $(Z, g : Z \to X)$ such that $fg = 0$, there exists a unique $h : Z \to X'$ such that the following diagram commutes $\ X' \xrightarrow{\iota} X \xrightarrow{f} Y$. The kernel is unique up to isomorphism.

$$X' \xrightarrow{\iota} X \xrightarrow{f} Y$$
$$h\uparrow \quad \nearrow g$$
$$Z$$

- A *cokernel* of a morphism $f : X \to Y$, denoted $\mathrm{coker}(f)$ is a pair (Y', σ) where $Y' \in \mathfrak{A}$ and $\sigma : Y \to Y'$, such that $\sigma f = 0$, and given another pair $(Z, g : Y \to Z)$

such that $gf = 0$, there exists a unique $h : Y' \to Z$ such that the following diagram

commutes $\quad X \xrightarrow{\ f\ } Y \xrightarrow[g]{\ \sigma\ } Y' \searrow_{\substack{\downarrow h \\ Z}}$. The cokernel is unique up to isomorphism.

- The *image* of a morphism $f : X \to Y$, denoted im (f), is the kernel (if it exists) of the cokernel of f.
- The *coimage* of a morphism $f : X \to Y$, denoted coim(f), is the cokernel (if it exists) of the kernel of f.

The proof of next proposition is left to the reader.

Proposition E.2.5 *Assume that the four objects (kernel, cokernel, image, coimage) exist for $f : X \to Y$. There is a unique morphism $\overline{f}$ such that the following diagram (called then the* canonical decomposition of a morphism*) is commutative:*

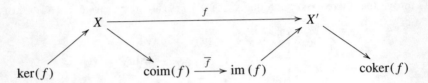

Abelian k–linear categories.

Definition E.2.6 A k-linear category $\mathfrak{A}$ is said to be *abelian* if the following conditions are satisfied.

(1) Every morphism in $\mathfrak{A}$ has a kernel and a cokernel.
(2) Every monomorphism is the arrow of a kernel and every epimorphism is the arrow of a cokernel.

By slight abuse of notation, from now on we shall often forget the "arrow part" of kernels and cokernels by mentioning only the "object part".

The canonical decomposition (see above Proposition E.2.5) provides an isomorphism in an abelian category, as shown by the next proposition (whose proof is left to the reader).

Proposition E.2.7

(1) *If $\mathfrak{A}$ is abelian, then for each morphism f, the corresponding morphism $\overline{f}$ is an isomorphism.*
(2) *In that case, the canonical decomposition of a morphism is unique up to a unique isomorphism.*

Exact sequences and exact functors.

Definition E.2.8 Let $\mathfrak{A}$ be an abelian category.

(1) A sequence of morphisms in $\mathfrak{A}$

$$X_1 \xrightarrow{\alpha_1} X_2 \xrightarrow{\alpha_2} \cdots \xrightarrow{\alpha_{n-1}} X_n$$

is said to be *exact* if, for all $i = 2, \ldots, n-1$, $\ker \alpha_i = \operatorname{im} \alpha_{i-1}$.

(2) A *short exact sequence* is an exact sequence of the shape

$$0 \longrightarrow X' \xrightarrow{\iota} X \xrightarrow{\pi} X'' \longrightarrow 0,$$

thus, we have

$$\begin{cases} \iota \text{ is a monomorphism,} \\ \pi \text{ is an epimorphism,} \\ \operatorname{im} \iota = \ker \pi. \end{cases}$$

Definition E.2.9 A functor $F : \mathfrak{A} \to \mathfrak{B}$ is *exact* if whenever $X' \to X \to X''$ is exact in $\mathfrak{A}$, then $F(X') \to F(X) \to F(X'')$ is exact in $\mathfrak{B}$.

Exercise E.2.10 Prove that a functor $F : \mathfrak{A} \to \mathfrak{B}$ is exact if and only if the image under F of any short exact sequence in $\mathfrak{A}$ is a short exact sequence in $\mathfrak{B}$.

Definition E.2.11 Let $\mathfrak{A}$ and $\mathfrak{B}$ be abelian categories. We say $\mathfrak{A}$ and $\mathfrak{B}$ are equivalent as abelian categories if there exist two exact functors $F : \mathfrak{A} \to \mathfrak{B}$ and $F' : \mathfrak{B} \to \mathfrak{A}$ such that $F \simeq F'$.

Exercise E.2.12 Let $\mathbf{vect}_k$ be the abelian k-linear category of finite dimensional k-vector spaces.

Let $\mathbf{Mat}_k$ be the category such that

- $\operatorname{Ob}(\mathbf{Mat}_k) = \mathbb{N}$
- For $m, n \in \mathbb{N}$, $\operatorname{Mor}(m, n)$ is the space $\operatorname{Mat}_{m,n}(k)$ of $m \times n$ matrices with entries in k, and where the composition of morphisms is given by matrix multiplication.

Prove that $\mathbf{vect}_k$ and $\mathbf{Mat}_k$ are equivalent as abelian categories.

Jordan–Hölder series and the Grothendieck group.

A nonzero object S in an abelian category $\mathfrak{A}$ is called *irreducible* if a monomorphism with goal S is either 0 or an isomorphism.

If $Y \hookrightarrow X$ is a monomorphism, by abuse of notation we denote by X/Y its cokernel.

- For X an object of $\mathfrak{A}$, we say that X *has finite length* if there exists a "filtration"

$$0 = X_0 \hookrightarrow X_1 \hookrightarrow \cdots \hookrightarrow X_n = X$$

such that X_i/X_{i-1} is irreducible for all $i = 1, \ldots, n$.

Such a filtration is called a *Jordan–Hölder series* of X.

• For S an irreducible object, its multiplicity in such a Jordan–Hölder series of X is the number of values of i for which X_i/X_{i-1} is isomorphic to S.

Theorem E.2.13 *If X has finite length and if S is an irreducible object, the multiplicities of S in any Jordan–Hölder series of X coincide.*

Such a multiplicity as in the above theorem is denoted by $m_S(X)$.

Definition E.2.14 Let $\mathfrak{A}$ be an abelian category all of whose objects have finite length. The Grothendieck group $\mathrm{Gr}(\mathfrak{A})$ is the free abelian group with basis the set $\mathrm{Irr}(\mathfrak{A})$ of isomorphism classes of irreducible objects in $\mathfrak{A}$.

For X an object of $\mathfrak{A}$, its class $[X]$ in $\mathrm{Gr}(\mathfrak{A})$ is defined by the formula

$$[X] := \sum_{S \in \mathrm{Irr}(\mathfrak{A})} m_S(X)[S].$$

If $0 \to X_0 \to \cdots \to X_n \to 0$ is an exact sequence in $\mathfrak{A}$, then

$$\sum_{i=0}^{n} (-1)^i [X_i] = 0.$$

Bibliography

[AB95]   J.L. Alperin, R.B. Bell, *Groups and Representations*, vol. 162 (Graduate Texts in Mathematics (Springer, Berlin, 1995)

[Aig79]   M. Aigner, *Combinatorial Theory*, vol. 234 (GTM (Springer, New York, 1979)

[Ben76]   M. Benard, Schur indices and splitting fields of the unitary reflection groups. J. Algebr. **38**, 318–342 (1976)

[Ben93]   D. Benson, *Polynomial Invariants of Finite Groups, London Mathematical Society Lecture Note Series* (Cambridge University Press, Cambridge, 1993)

[Bes97]   D. Bessis, Sur le corps de définition d'un groupe de réflexions complexe. Commun. Algebr. **25**, 2703–2716 (1997)

[Bou68]   N. Bourbaki, *Groupes et Algèbres de Lie, Chapitres IV, V, VI* (Hermann, Paris, 1968)

[Bou12]   N. Bourbaki, *Algèbre, Chapitre 8, Modules et Anneaux Semi-Simples* (Springer, Berlin, 2012)

[Bro10]   M. Broué, *Introduction to Complex Reflection Groups and Their Braid Groups*, vol. 1988 (LNM (Springer, Berlin, 2010)

[Bro13]   M. Broué, *Some Topics in Algebra, Mathematical Lectures from Peking University* (Springer, Berlin, 2013)

[CR87]   C.W. Curtis, I. Reiner, *Methods in Representation Theory I, II* (Wiley, New York, 1981/1987)

[Cur99]   C.W. Curtis, *Pioneers of Representation Theory*, vol. 15 (History of Mathematics (American Mathematical Society, Providence, 1999)

[DM85]   F. Digne, J. Michel, Fonctions $\mathcal{L}$ des variétés de Deligne–Lusztig et descente de Shintani, *Mémoires de la S.M.F.*, vol. 20 (Société Mathématique de France, 1985)

[Dri87]   V.G. Drinfeld, Quantum groups, in *Proceedings of the ICM Berkeley, 1986*, vol. 1 (Academic Press, London, 1987), pp. 798–820

[EGNO15]   P. Etingof, S. Gelaki, D. Nikshych, V. Ostrik, *Tensor Categories*, vol. 205 (Mathematical Survey and Monographs (American Mathematical Society, Providence, 2015)

[FKS81]   B. Fein, W. Kantor, M. Schacher, Relative brauer groups ii. J. Reine Angew. Math. **328**, 39–57 (1981)

[Gor80]   D. Gorenstein, *Finite Groups* (Chelsea, 1980)

[Hup67]   B. Huppert, *Endliche Gruppen I* (Springer, Berlin, 1967)

[Isa76]   I.M. Isaacs, *Character Theory of Finite Groups* (Academic Press, New York, 1976)

[Kas95]   C. Kassel, *Quantum Groups*, vol. 155 (Graduate Texts in Mathematics (Springer, Berlin, 1995)

[Lam98]   T.Y. Lam, Representations of finite groups: a hundred years, i. Not. Am. Math. Soc. **45**, 361–372 (1998)

© Springer Nature Singapore Pte Ltd. 2017

M. Broué, *On Characters of Finite Groups*, Mathematical Lectures from Peking University, https://doi.org/10.1007/978-981-10-6878-2

[Lan65]   S. Lang, *Algebra* (Addison-Wesley, New York, 1965)

[Lus87]   G. Lusztig, Leading coefficients of character values of Hecke algebras, in *Proceedings of Symposia in Pure Mathematics* (American Mathematical Society, Providence, 1987), pp. 235–262

[Mac71]   S. MacLane, *Categories for the Working Mathematician* (Springer (G.T.M.5), Berlin, 1971)

[MNO00]   G. Malle, G. Navarro, J.B. Olsson, Zeros of characters of finite groups. J. Group Theory **3**, 353–368 (2000)

[Ser12]   J.-P. Serre, *Linear Representations of Finite Groups*, vol. 42 (GTM (Springer, Berlin, 2012)

[ST54]   G.C. Shephard, J.A. Todd, Finite unitary reflection groups. Can. J. Math. **6**, 274–304 (1954)

[Ste03]   I. Stewart, *Galois Theory*, 3rd edn. (Chapman & Hall CRC Mathematics, Boca Raton, 2003)

[Ste12]   B. Steinberg, *Representation Theory of Finite Groups, An Introductory Approach, Universitext* (Springer, Berlin, 2012)

Index

A
(A, B)-bimodule, 17
Abelian k-linear category, 238
Absolutely irreducible character, 54
Absolutely irreducible module, 54
Algebraic integer, 225
$_A\mathbf{mod}$, 196
A-module-B, 17
Artin Theorem, 132
$\mathrm{Aut}(G)$, 22
$\mathrm{Aut}(X)$, 20

B
Big enough field, 133
g, x
$\mathrm{Cl}(\Omega)$, x
$[G/H]$, x
Braided category, 201

C
Canonical decomposition, 62
Category, 20
$\mathrm{CF}(G, k)$, 36
$\mathrm{CF}(G, \mathrm{Sec}_\pi^G(u), K)$, 141
$\mathrm{CF}_\pi(G, K)$, 141
$\mathrm{CF}(G, \mathbb{Q}_K^G)$, 48
$\mathrm{Cha}_K(G)$, 71
Character of the trivial representation, 44
Character table, 75
Characteristic degrees, 164
Characterization of reflection groups, 175
χ_M^a, 46
χ_G^{reg}, 52
χ_M^σ, 46
Class function, 36

D
Degree of a **Set**-representation, 26
Degree of a k-linear-representation, 29
$\deg(x)$, 155
Degrees, 164
Derived Subgroup, 34
Dihedral group, 19
Disjoint KG-modules, 124
$D_k G$, 180
Double centralizer property, 60
Double Cosets, 119
Doubly transitive, 59
Drinfeld Double, 195
Drinfeld element, 200
D_S, 49, 52

E
G, 72
e_G, 44

$\mathrm{Cl}(G)$, 36
$\mathrm{Cl}(G \times G)^{\mathrm{com}})$, 181
$C^{(n)}$, 72
Coevaluation map, 196
coev_M, 196
Cohen-Macaulay, 165
Commutator, 34
Complements of a Frobenius group, 148
Completely reducible representation, 43
Contragredient module, 197
Contragredient representation, 32
Control of the fusion of nilpotent π-subgroups, 141
Coxeter Group, 105
$\mathcal{CS}(G)$, 132

Printed in the United States
By Bookmasters